LETTERS TO PARENTS
in
Science

Anthony D. Fredericks

Illustrated by Dave Garbot

GoodYearBooks
An Imprint of ScottForesman
A Division of HarperCollins Publisher

Dedication

To Anita Meinbach, whose creative vision I admire and whose friendship I value . . . may they always be constants.

Cover photographs:
Globe and Astronaut: NASA
Butterfly: John Gerlach/Tom Stack & Associates
All other photographs: ScottForesman

GoodYearBooks

are available for most basic curriculum subjects plus many enrichment areas. For more GoodYearBooks, contact your local bookseller or educational dealer. For a complete catalog with information about other GoodYearBooks, please write:

GoodYearBooks
ScottForesman
1900 East Lake Avenue
Glenview, IL 60025

Book design by Street Level Studio.
Copyright ©1993 Anthony D. Fredericks.
All Rights Reserved.
Printed in the United States of America.

2 3 4 5 6 7 8 9 – PC – 01 00 99 98 97 96 95 94

ISBN 0-673-36079-2

Only portions of this book intended for classroom use may be reproduced without permission in writing from the publisher.

Preface

This book's purpose is to help you, the teacher, get parents actively involved in their children's science instruction. You may be asking, "With everything else I have to do during the school year, why should I communicate with parents?" The answer to that question, supported by years of field research, is clear. Numerous studies show that parents are the most important elements in a child's scholastic achievement and intellectual success. The support, encouragement, patience, and understanding of parents goes a long way in helping children grow cognitively as well as affectively. As much as 70 percent of a child's intellectual development takes place outside the classroom. Parents can help children appreciate science as an active subject—one in which people can share ideas and learn more about the world around them. Involving parents in school science programs has been proven to increase students' science awareness and appreciation. Science moves out from the pages of textbooks into the everyday lives of children. When parents are active participants in the science program, students have extra learning opportunities that go far beyond any teacher's manual or curriculum guide.

This book offers you a variety of tools through which you can employ parents as significant helpers in the encouragement and appreciation of science. Obviously, the intent is not to turn parents into scientists or students into geniuses. Rather, the idea is to provide some interesting and exciting ways in which your science program can be expanded and enlarged beyond the walls of your classroom and into the everyday lives of your students. Let's take a look at some of the critical factors that support a strong home/school relationship, particularly its impact on your science curriculum.

1. Teachers should help parents create regular, even daily, opportunities to share science books and science activities with their children.
2. Teachers need to let parents know that their participation and involvement will have important effects on their children's knowledge and appreciation of science.
3. Teachers and parents should promote science as a natural and normal part of everybody's lifestyle, not just something encountered in dry textbooks.
4. Regular and frequent communication between home and school can do more to ensure high levels of parent involvement than any other single factor.
5. The best kind of parental participation is that which lasts the whole year through—it involves a long-term commitment by teachers and families alike.

By following the guidelines above, you can "energize" your science program in fascinating ways. You can help your students appreciate the marvels and wonders of science far beyond the pages of your textbook. You can help them see the relevance of classroom projects and activities in a new and expanded light. You can demonstrate that science "happens" every day, every

From *Letters to Parents in Science* published by GoodYearBooks. Copyright ©1993 Anthony D. Fredericks.

hour, every minute, and that science is a process of sharing and discovering information with other people. The guidelines can help you greatly expand your teaching effectiveness.

This book provides a wealth of carefully prepared printed materials, such as letters, newsletters, and calendars, which you can duplicate and send home on a weekly or monthly basis. The reproducible materials contain direct and easy-to-implement activities and related children's literature. The materials can be incorporated regularly into the family's daily routines and schedules without turning the home into a "school away from school." The suggestions, ideas, projects, experiments, and reading lists are designed to encourage high levels of science competence and appreciation.

You will discover many benefits when you actively ask parents to be partners in the science education of their children. Parents will support and understand classroom efforts, because you have provided them with frequent information. Your students will discover an atmosphere of support and encouragement that can help them achieve positive attitudes toward and achievement in all dimensions of science. Students and their parents will see science as a relevant and pivotal component of every person's life.

What Children Need From Their Parents in Science

NEEDS	BENEFITS
Regular daily time	Children discover that science is a part of everyone's life every day.
Purposeful activities	Children learn that science is *active learning*, rather than the passive reading of textbooks.
Related to child's interests	There is always some aspect or area of science that can be coordinated with a child's interests. This personalizes science for each youngster.
Tolerance and patience	Parents and children discover that science is a constant process of learning and discovery—one in which all people have something to gain over time.
Support and encouragement	Children, like scientists, need the support of others and the stimulation that comes when ideas and possibilities are shared.
Informality	Much of what we know about the world of science comes through everyday experiences. Formal lesson plans are certainly not needed by parents.
Interaction	Science is best learned when it involves active communication between several parties. Science is not an isolated subject, but rather one made more "comfortable" when worked in teams.

From *Letters to Parents in Science* published by GoodYearBooks. Copyright ©1993 Anthony D. Fredericks.

The letters, newsletters, calendars, and other materials in this book are designed around the following five major subject areas:

1. **_Life Science_**: Where animals live. How animals grow and behave. The world of dinosaurs. How plants grow and reproduce. The importance of ecology. How to preserve the environment.
2. **_Physical Science_**: Simple machines and how they work. How heat, light, and sound are produced and transmitted. How to define and produce work, energy, and magnetism.
3. **_Earth Science_**: How volcanoes, earthquakes, weathering, and erosion change the earth. How oceans shape the earth and affect climate. Types of weather and the seasons of the year.
4. **_Space Science_**: The earth, sun, moon, and stars. The planets and the exploration of space.
5. **_The Human Body_**: Body support and movement. How the body's systems function. How to stay healthy. How the body changes throughout the human growth cycle.

Strong home-school bonds can help students achieve science comprehension and enlightenment. These materials, written in easy-to-understand language, help parents play active roles in the education of their children. They allow you to communicate to parents on a regular basis—not only about the dynamics of your classroom science program, but also about how their involvement can improve their children's science literacy.

These letters, newsletters, and calendars can be used throughout the year. They are suitable for any classroom or science curriculum. The book is a convenient, ready reference for any teacher wishing to expand students' science horizons through a positive partnership with parents. Your science program can become truly _energized_ when you enlist parents' aid and support.

From _Letters to Parents in Science_ published by GoodYearBooks. Copyright ©1993 Anthony D. Fredericks.

Contents

How to Use This Book   viii

Information Please

Introductory Letter
Evaluation Letter

Life Sciences

1. Animals: Growth
2. Animals: Behavior
3. Animals: Habitats
4. Animals: Categories
5. Dinosaurs: Big & Tall
6. Dinosaurs: Short & Small
7. Plants: Growth
8. Plants: Processes
9. Plants: Flowering
10. Plants and People
11. Ecology & Environment: Chains & Webs
12. Ecology & Environment: Preserving the Earth
13. Ecology & Environment: Conservation

Physical Sciences

14. Simple Machines
15. Heat
16. Electricity
17. Magnetism
18. Work and Energy
19. Light
20. Sound

Earth Sciences

21. Seasons
22. Water
23. Rocks and Soil
24. Changes in the Earth: Volcanoes & Earthquakes
25. Changes in the Earth: Weathering & Erosion
26. Oceans
27. Weather
28. Weather and Climate
29. Clouds and Storms

Space Sciences

30. The Earth
31. The Sun, Moon, and Stars
32. The Planets
33. The Solar System
34. Exploring Space

The Human Body

35. Body Support and Movement
36. Growing and Changing
37. Digestion and Circulation
38. The Brain and Sense Organs
39. Respiration and Excretion
40. Staying Healthy

Activity Calendars

Theme Newsletters

Growing Green: The World of Plants

Fur and Fins: The Lives of Animals

Living Things Need Each Other: Ecology and
    Environment

Things that Matter

Charge! Learning About Energy

The Real Dirt: Learning About the Earth

Weather or Not: The Forces of Weather and
    Climate

Beyond Earth: Exploring Planets and Space

The Body Human: Learning About Ourselves

Special Letters

A. Family Science Check-up

B. Can You Help Us?

C. Science Magazines for Children

D. Science Activity Books

E. Sources for Children's Literature in
    Science

F. Activities for Use with Science Books

G. Book-Sharing Questions

H. Science Supply Houses

I. Family Field Trip Sites

J. Things to Write For

K. Student Summary Sheet

L. Science Fair Timetable

M. Safety Rules

N. Potpourri

O. Ten Commitments for Parents

Certificate of Recognition

From *Letters to Parents in Science* published by GoodYearBooks. Copyright ©1993 Anthony D. Fredericks.

How to Use This Book

Each of the topics, projects, and activities in this book has been selected after a careful evaluation of current science textbooks, popular science books, and relevant theories and concepts about how children learn science. The letters are intended to encourage parents and children to work together in a relaxed and comfortable environment and learn about some of the marvels of science.

The core of this book consists of forty "Parent Letters," which are organized into five broad subject areas. The introductory section, "Information Please," includes an introductory letter and an evaluation letter. You may use the latter throughout the school year as a check on the effectiveness of the Parent Letters.

I encourage you to send home these letters to the parents of your students on a weekly basis, starting at the beginning of the school year. Here are some suggestions for distribution.

1. Remove the Introductory Letter from the book and sign your name in the appropriate space. Duplicate enough copies of the letter for the number of students in your class. You might have each child take a letter home to present to his or her parents, or hand a letter to each parent or guardian sometime during the first week of school (at "Back to School Night" or "Open House," for example). An alternate strategy would be to mail this initial letter home to parents. Make sure that each family receives the letter, and that you have the correct address to notify the parents about upcoming letters and activities.

2. Periodically, remove an additional letter from the book, sign your name in the appropriate space, duplicate it, and send it home. Please note that there is no specific sequence to these letters. You may use them in the order they are presented, or in an order in keeping with your science program.

3. Occasionally you might write appropriate notes or comments on the letters. This personalizes the letters even more and lets parents know that you are concerned and interested in their children's development.

4. I suggest that you establish one day of the week as "Letter Day." By sending the letters home on the same day each week, you notify both parents and children that the letters will be a regular part of your science curriculum and should be expected.

5. Other communications from the school or class to the parents should emphasize the importance of these letters. Continually remind parents about the letters. You might also point out that these letters are not intended as homework assignments, but rather as opportunities for all members of the family to work and play together in an atmosphere of support and encouragement.

6. Encourage parents to get in touch with you if they have any questions or concerns about any of the activities or projects mentioned in the letters. From time to time, invite parents to come to your classroom to visit and share some of *their* ideas and thoughts. You can also have parents share some of the activities and experiments they and their children work on together.

viii　　From *Letters to Parents in Science* published by GoodYearBooks. Copyright ©1993 Anthony D. Fredericks.

Alternate Distribution Ideas

The following are some additional strategies for getting these letters into the hands of parents:

A. Include a letter as part of a regularly distributed school or district newsletter.
B. Have students write letters home to parents about the value of these letters. The student letters can be photocopied and sent periodically throughout the year.
C. In all your contacts with parents—particularly those at the beginning of the year—be sure to mention the letters.
D. Ask if copies of the letters can be distributed through various community agencies or organizations (such as women's clubs, fraternal organizations, and churches and synagogues).
E. Staple the letters to homework assignments or activity sheets that you send home.
F. Obtain large sealable plastic freezer bags. Place a few items in each bag (magnifying lens, tweezers, etc.) along with a copy of a letter. The items in the bag are tools and materials completing an activity in the letter.
G. Periodically, write a brief note about each student's progress in science on selected letters. This can provide a most positive line of communication between home and school.
H. Periodically, call parents to remind them about the letters and suggest some extending or alternate activities to share with their children.
I. The principal should mention the letters in any communications he or she sends home to parents.
J. Encourage your students to prepare a special brochure that outlines the advantages of the letters and suggests ways they can be used at home. Distribute this brochure to parents as soon as possible after the beginning of the year.
K. Plan and schedule special workshops throughout the year. During these workshops, parents can learn about new activities and share some of their favorite ones with other parents.
L. Have your students help you put together a videotape on how a "typical" family uses one or more of the letters at home. Make copies of the videotape for classwide distribution.

I am certain that with a little creativity and imagination, you and your students will be able to devise alternate methods and procedures for getting these letters home. I have discovered that when students are involved as "recruiters" for their own parents, the success of these letters (and the improved levels of science competence and appreciation) are assured!

Special Features

In addition to these letters, this book contains several other features that will make your job easier. The first section, "Information Please," includes an introductory letter, which explains to parents the importance of their involvement in their child's science education. I encourage you to send this letter home during the first weeks of school.

An additional feature included in this first category is an evaluation letter, which is designed to enable parents to react to the letters that you have sent home. You should duplicate and send this letter home three or four times during the school year.

A third feature that has been incorporated throughout all the letters is a list of

From *Letters to Parents in Science* published by GoodYearBooks. Copyright ©1993 Anthony D. Fredericks.

relevant children's literature in science. Parents and children can take advantage of the resources of the school or local public library in broadening their awareness of and appreciation of specific science concepts.

An additional feature of each letter is the inclusion of activities, demonstrations, or experiments that parents and children can create and work on together. Each of these projects has been designed using common everyday household items. Each has been selected to broaden the student's understanding of a scientific principle.

Another feature of each letter is the inclusion of the "Fun Fact of the Week." This tidbit of information is designed to show parents and children some of the amazing wonders of the world around them. The intent is to stimulate interest in many aspects of the scientific world.

The reproducible activity calendars provide day-to-day year-round science adventures for the family. Each one can be duplicated and sent home with the students at the beginning of each month. You can send home the calendars for the summer months just before the end of the school year.

After the activity calendars, you'll find a collection of nine reproducible newsletters. These are home-publishing projects that families can complete and share together. Assembling the newsletters encourages family teamwork and makes direct connections between science and the family's life. These, too, can be duplicated and sent home each month of the school year.

The next feature of this book is the special letters section. These sheets allow you to share additional information and resources with parents. You can distribute the sheets separately from the regular letters or attach them to letters at intervals throughout the

year. No matter how you decide to use these special letters, you will find them to be important parts of your overall science curriculum.

The final feature of the book includes the award certificate. Distribute this certificate, signed by the appropriate officials (including you), at a special end-of-the-year ceremony.

You might find it appropriate to duplicate multiple copies of each letter (as well as the special letters) at the beginning of the school year. Arrange the duplicated letters in a three-ring binder according to the organization of your science textbook series. Then, at intervals throughout the year, you will be able to select materials quickly.

From *Letters to Parents in Science* published by GoodYearBooks. Copyright ©1993 Anthony D. Fredericks.

Information Please

Introductory Letter

Dear Parents:

Our class will be studying many new and exciting subjects in science this year. Your child will be learning about plants and animals, what makes up the universe, machines and the work they do, changes in the earth, and a host of other science facts and concepts. All of our lessons will be designed to help your child understand more about the world and make a host of self-initiated scientific discoveries.

I would like to invite you to become a partner in your child's science program this year. Your involvement will help your child attain higher levels of science competence and develop appropriate science skills that can last a lifetime. This partnership between school and home can provide your child with many extended opportunities to learn about the world.

In order to reinforce the work we are doing in the classroom, I will be regularly sending home prepared parent letters, calendars, and newsletters with activities for you and your child to share. These materials are designed to give you ideas that can help your child develop science skills. Each letter, calendar, and newsletter contains several choices of activities, experiments, and projects to share—activities that will reinforce your child's science education without disrupting your schedule. There are few materials to buy and few supplies needed—most are normally available in your own home. Your primary investment will be a few moments of your time each day. These few moments can make a world of difference in your child's education.

I look forward to your participation in our science program this year. If you have any questions about these materials or activities, please feel free to contact me. Let's work together to help your child succeed in science!

Sincerely,

Evaluation Letter #_____

Dear Parents,

For the past several weeks you and your child have shared the special parent letters sent home as part of our classroom science program. As you know, these letters are designed to help you help your child enjoy and learn science more! The intent is for both you and your child to become active participants throughout our entire science program.

I am very interested in learning about your reaction to the letters sent home during the past few weeks. Your input will help me design a science program that best meets the needs of your child as well as all students in the classroom. Would you please take a few moments of your time to complete the sections below? Please have your child return this completed form to school in the next few days. Thank you for your time and participation.

Sincerely,

My child and I found the science letters to be: (check one)
- _____ Very interesting
- _____ Good
- _____ Average
- _____ Not interesting

The part of the letters we enjoyed most was: (check one or more)
- _____ The background information on the different topics
- _____ The activities and projects
- _____ The "Fun Fact of the Week"
- _____ The suggested children's books
- _____ Other (please list)

We have obtained the suggested children's books (through our local public library or bookstore): (check one)
- _____ Many times
- _____ A few times
- _____ Seldom
- _____ Not at all

Comments:

I would like to receive additional science materials and activities to use at home.
- _____ Yes
- _____ No

Student's Name: _____

Parent's Signature: _____

Date: _____

Life Sciences

Animals: Growth

Dear Parents:

Children are fascinated by animals. In fact, of all the subjects we deal with in our science programs, animals continue to be among the most popular. And the growth and development of animals are topics that students continue to enjoy year after year.

Basically, there are two major groups of animals—**vertebrates** (animals with backbones), and **invertebrates** (animals without backbones). As you might imagine, these groups of animals grow and develop in different ways. Vertebrates are either born alive or hatched from eggs. Vertebrates move through several predictable stages until they reach adulthood. The frog, for example, begins life as an egg, becomes a tadpole, turns into a young frog, and then changes into an adult.

Invertebrates also pass through several stages in their growth. For example, a butterfly starts out as an egg, becomes a larva, turns into a pupa, and finally grows into an adult. The changes from one stage to another are usually more dramatic with invertebrates than they are with vertebrates.

You can help your child appreciate and learn more about the growth of animals with one or two of the following activities. Be sure to share the marvelous world of animals with your child throughout the entire year.

1. If possible, provide your child with a small pet such as a tropical fish, mouse, hamster, or bird. Have your child take photographs of the pet once a week for several months. Have your child arrange the photos in an album with an appropriate label for each one. Talk with your child about the growth processes the pet undergoes throughout its life.

2. Obtain the bones or skeletons of several different animals (for example, a fish, a chicken, a plastic model of a human skeleton). Talk with your child about the importance of the skeleton in the growth process. What similarities or differences does your child note among the skeletons?

3. Have your child look through several old magazines for pictures of animals (both vertebrates and invertebrates). Have your child cut out examples of young animals and adult animals and create a colorful collage of the animal kingdom. Talk about some of the similarities and differences between adults and their children (among animals as well as humans).

Fun Fact of the Week

The Gippsland Gurgling Earthworm from Australia often reaches a length of 12 feet. It is one of the world's largest invertebrates.

Suggested Children's Books

Goor, Ron and Nancy. *Insect Metamorphosis: From Egg to Adult.* New York: Atheneum, 1990.
Henley, Karyn. *Hatch!* Minneapolis, MN: Carolrhoda, 1980.
Kaufman, Joe. *Wings, Paws, Hoofs, and Flippers: A Book About Animals.* New York: Golden, 1981.
Lacey, Elizabeth. *The Complete Frog: A Guide for the Very Young Naturalist.* New York: Lothrop, 1989.

Sincerely,

Life Sciences

Animals: Behavior

Dear Parents:

Do you have a pet at home? Have you ever noticed that it behaves in certain ways? Your cat expects food whenever you turn on the can opener. Your dog makes a certain sound whenever it wants to be let outside. Your hamster sleeps during the day and plays at night. Some of the behaviors that animals have are learned. Others are based on instinct (they are born with instinctive behaviors). For example, kittens learn to catch mice because they watch their parents do it (learned behavior). On the other hand, most young animals know how to suckle milk from their mothers (instinctive behavior).

The ways animals behave are sometimes mysterious. One of the most common animal behaviors is the need to live together in groups. Also, animals provide differing levels of care for their young. Animals must continually learn new skills throughout their lives and balance those with the instinctive skills with which they are born.

Looking into the diverse ways in which animals behave is an area ripe for investigation and discovery. Students soon learn that there is much to know about animal behavior and that it often provides insights into human behavior. One or two of the following activities might be appropriate to share with your child.

1. Ask your child to record and chart some of the distinctive behaviors of the family pet. Have your child observe other similar pets in the neighborhood to determine their distinctive behaviors. For example, do all dogs tend to do the same kinds of things? How does the behavior of poodles differ from that of retrievers? In what ways does the behavior of dogs differ from that of cats? Hamsters? Fish? Be sure to discuss some of your child's findings.

2. Make an ant farm. Obtain a large glass jar. Fill it about three-fourths full with a mixture of soil and sand. Place a small moist sponge and a sugar cube in the jar. Hunt throughout a nearby wooded area for an ant colony. Place the jar on its side in the middle of the ant swarm. Use a stick to push a few ants into the jar. Other ants will soon join the ants in the jar. Screw on the lid very tightly. Wrap the outside of the jar with black paper. After about 10 days take off the paper and ask your child to note what the ants have done inside the jar (you may need to occasionally moisten the sponge and replenish the sugar cube).

3. Obtain an inexpensive bird feeder from a local hobby store, pet shop, or hardware store. Fill it with birdseed and place it outside a window. Have your child observe, not only the various types of birds that visit the feeder, but what the birds do while at the feeder. Do some types of birds behave differently than others? Are there quiet eaters? Are there noisy eaters? Have your child keep track of the birds' behavior for an extended period of time.

Fun Fact of the Week

The hummingbird and the Abyssinian Blue Goose are the only birds that can fly backwards.

Suggested Children's Books

Pope, Joyce. *Do Animals Dream? Children's Questions About Animals Most Often Asked of the Natural History Museum.* New York: Viking, 1986.

Powzyk, Joyce. *Animal Camouflage: A Closer Look.* New York: Bradbury, 1990.

Selsam, Millicent. *Where Do They Go? Insects in Winter.* New York: Four Winds, 1982.

Sussman, Susan and Robert James. *Lies (People Believe) About Animals.* New York: Whitman, 1987.

Sincerely,

Life Sciences

Animals: Habitats

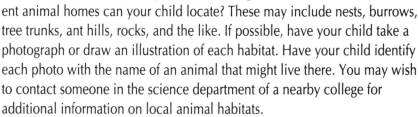

Dear Parents:

Have you ever thought about the different places animals live? Animals inhabit burrows, caves, nests, pens, rivers, and webs. Animals live in the highest trees and the deepest oceans. They live in the coldest regions and the most tropical areas of the world. No matter where you journey in the world, there will be several types of animal habitats.

Usually animals live in a particular place for one of two reasons—protection or survival. Many birds live in nests in the tops of trees to stay away from predators on the ground. Ants live in colonies underground to protect themselves from larger predators and from the elements. The proximity of food for survival also determines where animals live. Monkeys, for example live in the branches of trees so that they can easily obtain the fruits and other foods they need. Whales live near the surface of the ocean so that they can obtain the plankton and other organisms they need to eat. The habitats in which animals live are the result of a process known as **adaptation**—functional and behavioral changes that help an organism survive and reproduce. In other words, if a habitat did not allow an organism to obtain food or protect itself, then that organism would probably die out (that is, unless it found another home).

Help your child appreciate and learn more about the habitats of animals by selecting one or two of the following activities. Keep in mind that the information shared will most likely pertain to animals in their natural environments — rather than domesticated animals such as dogs.

1. Take your child on a walking "field trip" of your town or neighborhood. How many different animal homes can your child locate? These may include nests, burrows, tree trunks, ant hills, rocks, and the like. If possible, have your child take a photograph or draw an illustration of each habitat. Have your child identify each photo with the name of an animal that might live there. You may wish to contact someone in the science department of a nearby college for additional information on local animal habitats.

2. You and your child may enjoy constructing or obtaining one or more different animal habitats, which you can place outside your home or keep indoors for daily observation. Bird feeders and bird houses, bug houses, aquariums, terrariums, and the like would all be appropriate. Many larger toy stores sell a variety of animal habitats. You can also order animal homes from three excellent mail order houses, including: Aquarium and Science Supply Co. (101 Old York Rd., Jenkintown, PA

19046); Insect Lore Products (P.O. Box 1535, Shafter, CA 93263); and Delta Education (P.O. Box 950, Hudson, NH 03051).

3. With your child, talk about the habitats of animals in zoos. Are they appropriate? How are zoo habitats different from the habitats of animals in the wild? If possible, visit a local zoo and help your child make a list of the major differences between life in the zoo and life in the wild. How would your child feel about living in a zoo environment? How would it compare to the habitat he or she currently lives in?

Fun Fact of the Week
The silk in a spider's web is five times stronger than an equivalent filament of steel.

Suggested Children's Books
Arnold, Caroline. *Five Nests.* New York: Dutton, 1980.
Banks, Merry. *Animals of the Night.* New York: Scribner's, 1990.
George, William and Lindsay. *Beaver at Long Pond.* New York: Greenwillow, 1988.
Milne, Larus and Margery. *Gadabouts and Stick-at-Homes: Wild Animals and Their Habitats.*
New York: Sierra/Scribner's, 1980.

Sincerely,

Life Sciences

Animals: Categories

Dear Parents:

Animals are certainly part of our everyday lives. Whether we have pets at home, visit zoos or wild animal preserves, or watch birds on their annual migration patterns, we are fascinated with the usual and unusual members of the animal kingdom.

Animals are divided into two main groups—**invertebrates**, or those animals without backbones, and **vertebrates**, animals with backbones. Invertebrates constitute about 96 percent of all the animal species on the earth. Within the invertebrates group there are several subgroups of animals. These include sponges, animals with stinging cells (coral, jellyfish), worms, spiny-skinned animals (starfish, sea urchin), and soft-bodied animals (clam, octopus). Another group of invertebrates includes the arthropods (crabs, spiders, insects, centipedes). The arthropods outnumber all other animal species combined. (Note: Students often confuse spiders and insects. A spider has two body parts and eight legs, while an insect has three body parts and six legs.)

Vertebrates, too, are divided into different groups. These include birds (duck, penguin); mammals (cat, squirrel); fish (shark, goldfish); amphibians (frog, salamander); and reptiles (turtle, lizard). (Note: Children sometimes confuse amphibians and reptiles. An amphibian begins its life in water and breathes with gills. Later, an amphibian grows lungs and move to land. A reptile breathes with lungs all through its life.)

The world of animals is filled with many amazing creatures. You can help your child understand and appreciate the variety of animals by selecting one or two of the following activities. Be sure to involve other family members whenever possible.

1. You and your child can grow your own frogs. A kit is available from Holcombs Educational Materials (3205 Harvard Ave., Cleveland, OH 44105). Kit #998-0125H includes a container, food, instructions, and a coupon for live tadpoles. Or check local toy stores or teacher supply stores in your area for similar kits.

2. Have your child make a chart that has been divided into two sections— "Invertebrates" and "Vertebrates." Take a walking "field trip" with your child though your neighborhood or community. Have your child record all the animals you see during the walk. Look inside trees, up on electrical wires, under rocks, and behind buildings.

Upon your return home, discuss whether you found more vertebrates or invertebrates on the journey. What types of animals seem to predominate?

3. The biology departments of most colleges usually have collections and displays of various types of animals. Contact a nearby college or university and ask if you and your child can visit any displays or exhibits. Larger universities often have museums and other special buildings that house extensive displays of many different kinds of animals. Your child might want to bring a camera to take photographs (check with the museum before bringing in the camera). The photos can be assembled into albums or notebooks.

Fun Fact of the Week
Frogs must close their eyes in order to swallow.

Suggested Children's Books
Arnold, Caroline. *Animals That Migrate.* Minneapolis: Carolrhoda, 1982.

Johnson, Ginny and Judy Cutchins. *Scaly Babies: Reptiles Growing Up.* New York: Morrow, 1988.

Settle, Joanne and Nancy Baggett. *How Do Ants Know When You're Having a Picnic? (And Other Questions Kids Ask About Insects and Other Crawly Things).* New York: Atheneum, 1986.

Simon, Seymour. *101 Questions and Answers About Dangerous Animals.* New York: Macmillan, 1985.

Sincerely,

Life Sciences

Dinosaurs: Big and Tall

Dear Parents:

Ask any group of 100 kids what their favorite topic in science is and 99 of them will tell you, "dinosaurs." Dinosaurs have always been one of the most fascinating areas of science for both children and adults. Filled with mystery and adventure, the lives of dinosaurs have been a never-ending source of amazement and speculation for kids of all ages.

Even though humans have been intrigued by dinosaurs for some time, this area of science is often filled with many misconceptions. Motion pictures to the contrary, no human being has ever seen a live dinosaur. In fact, dinosaurs died out about 60 million years before humans appeared on earth. Also, many youngsters believe that all dinosaurs were huge lumbering creatures. In fact, most of the dinosaurs we know of were about the size of chickens.

You can open up new worlds of discovery and imagination when you share activities and books about dinosaurs with your child. The suggestions below are just some of the ways your family can learn about dinosaurs together.

1. You and your child might enjoy renting some dinosaur-related movies from your local video store. As you watch the movies, make a list of some of the physical characteristics of dinosaurs portrayed on the screen. Afterwards, compare your list with the information supplied in the books at the end of this letter. What major differences do you note?

2. Using the illustrations from several books, you and your child can create the head of a large dinosaur. Obtain some papier-mâché from a local hobby store and follow the directions for mixing. After your dinosaur head is complete, be sure to paint it. (Here's another interesting fact: although most dinosaurs are illustrated in greens and browns, scientists are not sure what color they were.)

3. You and your child might enjoy making a wall mural of a dinosaur scene. Obtain some newsprint from your local newspaper or hobby store. Work with your child to create an oversize painting. What other types of living things should be included in the mural?

Fun Fact of the Week

The largest known dinosaur was Ultrasaurus. It was 15 times larger than an African elephant, taller than a five-story building, and had legs more than twenty feet long.

Suggested Children's Books

Cohen, Daniel. *Monster Dinosaur.* New York: Lippincott, 1983.

Jacobs, Francine. *Supersaurus.* New York: Putnam, 1982.

Lauber, Patricia. *The News About Dinosaurs.* New York: Bradbury, 1989.

Robinson, Howard. *Ranger Rick's Dinosaur Book.* Washington, DC: National Wildlife Federation, 1984.

Sincerely,

Life Sciences

Dinosaurs: Short and Small

Dear Parents:

It's a fact—kids love dinosaurs! Indeed, so do adults. Exploring the world of dinosaurs is often a child's first look into the world of science—a look that can last far beyond the years in the classroom. Visit any bookstore or library and you will be amazed at the number of books on dinosaurs. There are books about how dinosaurs lived, what they ate, how they raised their young, and how they were wiped off the face of the earth.

One fact that amazes many youngsters (and many adults) is that most dinosaurs were small—often the size of dogs. While we are used to illustrations (and movies) that portray dinosaurs as huge, lumbering creatures, the average dinosaur could have easily fit in the trunk of a car (that is, if cars have been around when dinosaurs were).

Families will find the subject of dinosaurs ripe for all kinds of fascinating discoveries. These discoveries not only form the basis for future explorations into the world of science, but they also help illustrate that people of all ages can make science an interesting part of their everyday lives.

Select one or two of the following activities to share with your child. You will probably discover other activities for you and your child to do together as you read and learn more about the dinosaurs' world.

1. Have your child read several different books of dinosaurs (use the list below to start). Ask your child to record the lengths of the various types of dinosaurs mentioned in the books. Provide your child with a spool of string or twine. Have your child measure lengths of the string according to the different lengths (or heights) of various dinosaurs. If possible place these pieces of strings in your yard, driveway, or sidewalk to see how the different lengths compare. You might also help your child use graph paper to make the comparisons.

2. After your child has read several books about dinosaurs, have him or her create menus of what a dinosaur might eat during the course of a day. Keep in mind that most dinosaurs were **herbivores** (plant eaters), while only a few were **carnivores** (meat eaters). Discuss the kinds of foods a dinosaur might enjoy munching on during the day.

3. Obtain some pipe cleaners from a local hobby or tobacco shop. Ask your child to create several different dinosaur skeletons based on illustrations in books. Ask your child to explain the features included in the dinosaurs and then to set up a special "dinosaur museum" exhibit in the family room or kitchen.

Fun Fact of the Week

The closest living relatives of dinosaurs are birds. Some scientists believe that bird feathers evolved from dinosaur scales.

Suggested Children's Books

Barton, Byron. *Dinosaurs, Dinosaurs.* New York: Crowell, 1989.
Freedman, Russell. *Dinosaurs and Their Young.* New York: Holiday, 1963.
Sattler, Helen. *Baby Dinosaurs.* New York: Lothrop, 1984.
Simon, Seymour. *The Smallest Dinosaurs.* New York: Crown, 1982.

Sincerely,

Life Sciences

Plants: Growth

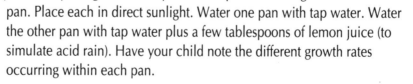

Dear Parents,

Whether you live in the city or the country, in the mountains or along the seashore, plants are everywhere. Plants can be found near the Arctic Circle and in the harshest deserts. Plants display a wide variety of colors (from white to black) and sizes (from the tallest sequoia to the smallest fungi). In fact, plants can be found in almost every location on the planet and in every size, shape, and color.

It's important for students to understand the wide variety of plants in the world. It's equally important for youngsters to understand how plants grow and mature. Some plants grow very rapidly. Some varieties of bamboo, for example, can grow up to three feet in one day. Other plants grow very slowly (the sequoia takes nearly 100 years to reach maturity. Some plants, such as a Bristlecone Pine tree in western Nevada, grow to be more than 4,000 years old. There are a wide variety of activities you and your child can participate in together to enjoy the magical world of plants and their different growing patterns. Try one or two of the following:

1. Fill several clear plastic cups with potting soil. In each one, plant several seeds from one type of vegetable (beans, radish, and squash work well). Place the seeds around the sides of the plastic cup (so they may be seen from the outside). Have your child place the cups in direct sunlight and water them occasionally. As the seeds begin to sprout, have your child note the growth patterns of the various plants over a period of several days.

2. Fill two baking pans with potting soil. Have your child sprinkle some grass seed over each pan. Place each in direct sunlight. Water one pan with tap water. Water the other pan with tap water plus a few tablespoons of lemon juice (to simulate acid rain). Have your child note the different growth rates occurring within each pan.

3. Obtain two clean yogurt containers (with lids). Have your child fill each to the top with bean seeds. Have your child pour as much water as possible into one of the cups. Cover each container with a lid and place both of them in a warm place. After several hours (or a full day) have your child note what happens to the container that had seeds and water. Why did the lid pop off this container? Have your child examine the seeds in both containers to arrive at a possible answer.

Fun Fact of the Week

The giant sequoia tree may take up to 175 years to begin to flower.

Suggested Children's Books

Budlong, Ware and Mark Fleitzer. *Experimenting with Seeds and Plants.* New York: Putnam, 1970.

Burnie, David. *Plant.* New York: Knopf, 1989.

Gutnik, Martin. *How Plants Make Food.* Chicago: Childrens Press, 1976.

Heller, Ruth. *Plants That Never Bloom.* New York: Grosset and Dunlap, 1984.

Tinsley, Thomas. *Plants Grow.* New York: Putnam, 1971.

Sincerely,

Life Sciences

Plants: Processes

Dear Parents:

When we walk through the forest, buy flowers at florist shops, or plant vegetables in our gardens, we often don't think deeply about the complex organisms called plants. Plants to us are simple things that sprout from seeds, stand up in the sun, grow to a predictable height, and eventually die or are cut down. Yet, inside every plant, a very complicated set of processes is taking place—a series of events that rivals some of the processes of more advanced forms of life.

The plant process most of us are familiar with is **photosynthesis**, the means whereby a plant makes it own food. It does this by trapping sunlight and changing it to a form of chemical energy. That chemical energy is combined with carbon dioxide in the air to make a special kind of sugar which is transported to all parts of the plant. Much of the oxygen we breathe is released by plants through photosynthesis.

Transpiration is the process by which a plant transports water throughout its system. Plants typically lose water through cells in their leaves. This loss of water pulls up more water from the roots (a corn plant can transpire about two quarts of water a day). **Respiration** is the process whereby plants use oxygen to break down sugar into carbon dioxide and water. This action also releases energy from the sugar—energy the plant's cells need to do their work. Respiration is a process shared by plants and animals. Another process of plants is **fertilization**—when a **sperm cell** joins with an **egg cell** to form a seed. Each seed has a **seed coat**, an **embryo**, and stored food. Some plants may produce millions of seeds in a year; others produce only one or two. The largest known seed is the double coconut which can weigh as much as 60 pounds.

Children are often amazed to discover all the complexities of plants. They are surprised to find out about the many processes that continuously take place in the plants around their homes or in their gardens. You can help your child learn more about these processes by sharing one or two of the following activities.

1. Obtain two similar potted plants of equal vigor and height. Have your child select twenty random leaves on one plant. Ask your child to smear a thin layer of petroleum jelly on the tops of ten leaves and the bottoms of ten other leaves (on the same plant). Place both plants on a window sill and water and fertilize as necessary. After several days have your child observe the two plants and note any differences. What happened to the leaves with petroleum jelly on the top? What happened to the leaves with petroleum jelly on the bottom? What can this tell you about some of the processes of plants?

2. Obtain two similar potted plants (of equal vigor and height). Water and fertilize the plants as necessary. Have your child place a large clear cellophane bag over the top of one plant, securing it around the pot with a rubber band. The bag should be as airtight as possible around the plant. Place the plants on a window sill. After several days have your child note what happens inside the bagged plant. Your child may wish to occasionally (every 2 or 3 days) remove the bag from that plant and measure the amount of water that has been transpired by the plant. Be sure to keep both plants adequately watered.

3. Have your child conduct some library research to determine how seeds travel. Have your child construct a large chart with the words "Water," "Wind," "Animals," and "Other" across the top. Ask your child to record the names of plants in one or more of the categories on the chart. He or she might want to look at plants such as the milkweed, dandelion, coconut, thistle, and sugar maple. What devices do some seeds have that help transport them to new areas?

Fun Fact of the Week
In tropical rain forests, plants known as **epiphytes** grow on the highest branches of trees. These plants have no roots, yet obtain water and minerals directly from the humid air.

Suggested Children's Books
Heller, Ruth. *Plants That Never Bloom.* New York: Grosset, 1984.
Johnson, Sylvia. *How Leaves Change.* Minneapolis: Lerner, 1986.
Nussbaum, Hedda. *Plants Do Amazing Things.* New York: Random House, 1977.
Overbeck, Cynthia. *Carnivorous Plants.* Minneapolis: Lerner, 1982.

Sincerely,

Life Sciences

Plants: Flowering

Dear Parents:

Flowers are an important part of our world. They decorate our homes and businesses, provide beauty throughout our cities and towns, and fill our lives with color. In fact, one of the most recognizable features of the world of nature is the abundance and diversity of flowers.

Flowers are also important for another reason. They are the chief way in which many plants are able to reproduce. Not all plants produce flowers; however, the largest category of plants in the world are those that use flowers to reproduce. Some plants, such as grasses, have flowers that are almost too small to be seen. Other plants, such as orchids, have large, bright, and colorful flowers. Flowers are of two types. **Perfect flowers** have both male and female reproductive organs. Peas and apples are two examples. **Imperfect flowers** contain only male or female reproductive organs. Two plants with imperfect flowers are maple trees and corn. Sexual reproduction by flowers requires **pollination**, or the transfer of pollen from one flower to another. Pollination is usually the work of animals or wind. In hybrid corn and other crops, the plants are pollinated artificially.

You can help your child appreciate the beauty and wonder of flowering plants with one or two of the following activities. Take time to discuss how some of these ideas relate to the flowers in or around your home.

1. Obtain a large wall map of the United States (available from many stationery, office supply, or toy stores). Post the map in your child's room. Ask your child to locate the state flower for each of the fifty states. Have your child pin the name of each flower along with an accompanying picture, photograph, or illustration near the outline of each state. This activity would be ideal for all family members to share together.

2. Obtain several lima beans and soak them in water overnight. Very carefully, cut several beans in half, lengthwise. Have your child identify the parts of the bean seed (seed coat, embryo, stored food). Discuss with your child the fact that each seed contains a very small plant (the embryo), which will use the stored food to begin its growth process. This process is typical of most flowering plants.

3. Obtain several different kinds of flowers from your garden, a local nursery, or florist. Have your child examine each for similarities and differences. Are the flowers more alike than different? Other than color, size, and shape, what are some of the major differences? Later, work with your child to create illustrations of each flower. For each, label the major parts (sepals, petals, stamens, pistils).

Fun Fact of the Week
There are more than 250,000 species of flowering plants in the world.

Suggested Children's Books
Crowell, Robert. *The Lore and Legend of Flowers.* New York: Crowell, 1982.
Ehlert, Lois. *Planting a Rainbow.* San Diego, CA: Harcourt, 1988.
Robbins, Ken. *A Flower Grows.* New York: Dial, 1990.
Selsam, Millicent. *Tree Flowers.* New York: Morrow, 1984.

Sincerely,

Life Sciences

Plants and People

Dear Parents:

Children are often amazed to discover that without plants, human beings could not survive. We depend directly on plants for much of our food (cereals, tomatoes, nuts, apples, tea), and indirectly for the rest of what we eat (meat, milk, eggs). Plants also provide much of our clothing materials (cotton, linen, rayon, flax). Plants are a chief source of many of the medicines we use (penicillin, digitalis). We also use plants to build and decorate our homes (lumber, paint).

Just as we are dependent upon plants for our daily lives, plants are dependent upon us for their care and maintenance. Vegetables and fruits must be cultivated and fertilized. Seeds must be gathered and preserved for annual plantings. Endangered species must be preserved for future generations.

Since the lives of humans and plants are intertwined, it is important for youngsters to develop an appreciation for all the products we get from plants. Without our care and maintenance, many plant species will die out, never to return. Without constant attention, many of the food sources in the world will be threatened and millions of people will face starvation.

Helping your child understand his or her role in preserving plants of all kinds for future generations is an important job. You and your child may wish to select one or two of the activities that follow. As you pursue these activities, be sure to include other family members in your discoveries and discussions.

1. Have your child make a large chart with the words "Food," "Clothing," "Shelter," and "Other" printed across the top. Invite your child to take a "field trip" through the house and record the names of items made from plants in each of the proper categories. How much of the living room comes from plant products? What would the living room look like if every plant-related product were removed? How many plant-related products are in your child's room?

2. Discuss with your child ways in which humans alter, pollute, or destroy plant **habitats** (places where plants live). Discuss some actions humans do intentionally (cutting down large areas of forest land) or unintentionally (forest fires) that affect whether plants can or will live in a particular area.

3. Encourage your child to contact a local Sierra Club chapter, National Wildlife Federation club, or branch of the National Audubon Society to find out what they are doing to preserve some of the plant life in your area. You can write to the national headquarters of each organization to ask about the nearest local chapter or branch: Sierra Club (730 Polk St., San Francisco, CA 94009); National Wildlife Federation (1412 16th St. NW, Washington, DC 20036); National Audubon Society (645 Pennsylvania Ave. SE, Washington, DC 20003).

Fun Fact of the Week

It takes 63,000 trees to produce the newsprint for one Sunday edition of the *New York Times.*

Suggested Children's Books

Brown, Marc. *Your First Garden Book.* Boston: Little, Brown, 1981.
Kuhn, Dwight. *More Than Just a Vegetable Garden.* New York: Silver Press, 1990.
Oechsli, Helen and Kelly Oechsli. *In My Garden: A Child's Gardening Book.* New York: Macmillan, 1985.
Selsam, Millicent. *Play with Plants.* New York: Morrow, 1978.

Sincerely,

Life Sciences

Ecology and Environment: Chains and Webs

Dear Parents:

All of the living things that live together in one place make up a **community**. The members of that community, whether they are plants or animals, are all dependent upon one another for survival. Some animals eat plants, some plants live off of other plants, and some animals eat other animals. All living things are part of a food chain—that is, energy and materials are passed through a line from one living thing to another.

An example of a food chain is when a grasshopper eats a blade of grass, a frog eats the grasshopper, a snake eats the frog, and an eagle eats the snake. However, nature is not quite that simple. That's because there may be many other kinds of animals that could eat the grass, or eat the grasshopper, or even eat the snake. When many living things are involved—that is, when several food chains are intertwined—that is called a **food web**. Food webs occur all the time in nature. Human beings are part of a very complex food web that includes large numbers and varieties of plant and animal life.

Students need to understand how we humans need all kinds of plants and animals for survival. The other living things in our world also depend on one another. You may wish to select one or two of the following activities to help your child understand food chains and food webs.

1. You and your child can write a letter to the National Wildlife Federation (1412 16th Street NW, Washington, DC 20036). Ask for information and printed materials on various types of food webs.

2. At the top of a large sheet of paper write the words "Herbivore," "Carnivore, " and "Omnivore." Help your child find the definitions in a dictionary. Then have your child look through several old magazines and cut out pictures of animals. Glue the animal pictures into the sections on the paper. Take time to discuss reasons why your child has placed certain animals in a particular category.

3. All food chains begin with the sun as the primary source of energy. Have your child draw a series of illustrations (with arrows in between) showing the progression of energy from the sun through a plant and continuing through several different animals. Be sure to provide your child with an opportunity to discuss his or her chain with other members of the family.

4. What did you have for dinner last night? Choose that meal or something else you ate this week. Diagram how your dinner fits into a food chain.

Fun Fact of the Week

Scientists have estimated that there are approximately 30 million species of living things in the world. Scientists have also calculated that about 150,000 species per year are now becoming extinct.

Suggested Children's Books

Cole, Sheila. *When the Tide is Low.* New York: Lothrop, 1985.
Dekkers, Midas. *The Nature Book: Discovering, Exploring, Observing, Experimenting with Plants and Animals at Home and Outdoors.* New York: Macmillan, 1988.
Hughey, Pat. *Scavengers and Decomposers: The Cleanup Crew.* New York: Atheneum, 1984.
Schwartz, David. *The Hidden Life of the Meadow.* New York: Crown, 1988.

Sincerely,

Life Sciences

Ecology and Environment: Preserving the Earth

Dear Parents:

Here are a few statistics to think about and discuss with family members.

The average American uses the equivalent of seven trees every year.
The energy saved from recycling one glass bottle will light a 100-watt bulb for four hours.
Over a billion trees are used to make disposable diapers every year.
An estimated 14 billion pounds of trash are dumped into the sea every year.
Every week, about 20 kinds of plants and animals in the world become extinct.

If those facts and figures shock you, consider that they are only the tip of the environmental iceberg. Many scientists believe that the earth is in peril simply because of what we, as humans, are doing on a daily basis. That means that our daily habits and everyday actions are having a profound and significant impact on the balance of nature. How we live today also affects how our children will be able to live in the future. The resources we use, the ways in which we care for our land, and our treatment of wildlife will all have a direct impact on the quality of life inherited by our children.

It seems safe to say that students today need to take an active interest in efforts for preserving the earth. Making them aware of the statistics above is one step. Children also need to be aware of the opportunities they have for ensuring and preserving a quality of life that is threatened and endangered. You and your child might select one or two of the suggestions below to get started.

1. You and your family can "adopt" an animal. Through the "adoption" process your family will receive a photo and fact sheet about your animal. The money you send in will be used to care for and feed it. Your family might want to consider caring for an endangered species. For more information on animal adoption, contact the American Association of Zoological Parks and Aquariums (4550 Montgomery Ave., Suite 940N, Bethesda, MD 20814).

2. Preserving water sources and resources can be an important part of family activities. This means not only cutting down on the amount of water used in the home, but also making sure that nearby rivers and streams do not become polluted with trash and wastes. You and your child can obtain a free booklet entitled "Save Our Streams" by writing to The Izaak Walton League of America (1401 Wilson Blvd., Level B, Arlington, VA 22209).

3. Your family can start your own recycling program. Perhaps you and your child would like to find out about ways in which kids can make a difference in preserving the environment. Or maybe your family would like to get some news about environmental "success stories" from around the country. If so, you and your child can write to Renew America (Suite 710, 1400 16th St. NW, Washington, DC 20036) for information and details.

Fun Fact of the Week
Americans generate 160 million tons of rubbish each year—that's 3-1/2 pounds a day for every man, woman, and child.

Suggested Children's Books
Baker, Jeannie. *Where the Forest Meets the Sea.* New York: Greenwillow, 1988.
Cherry, Lynn. *The Great Kapok Tree: A Tale of the Amazon Rain Forest.* New York: Gulliver, 1990.
National Wildlife Federation Staff. *Endangered Animals.* Washington, DC: National Wildlife Federation, 1989.
Pringle, Lawrence. *Saving Our Wildlife.* Hillside, NJ: Enslow, 1990.

Sincerely,

Life Sciences

Ecology and Environment: Conservation

Dear Parents:

How we care for the earth today will have a tremendous impact on the kind of planet our children will inherit in future years. Our care of the earth's inhabitants—both plant and animal life—will determine, in large measure, the kinds of foods, recreation, and populations that are available twenty, fifty, or a hundred years from now. Helping our children become conservators of the earth is an important goal of any science program and can be an important goal of every family, too.

Conservation, however, is more than just recycling our aluminum cans and cutting down on our daily water use. It is a conscious effort by everyone to save and preserve all of our precious resources so that they may be available for generations to come. That effort is not easy. It takes lots of planning, lots of time and trouble, and lots of people working together to ensure a high quality of living for ourselves and our biological neighbors. There are many organizations that can help in providing you and your family with information and ideas on how to conserve. Write or call them for materials (many of which are free). Some possibilities include the following:

For a free brochure on recycling, write to the Environmental Defense Fund (257 Park Ave. S, New York, NY 10010).

For information on planting trees and celebrating Arbor Day, write to the National Arbor Day Foundation (Arbor Lodge 100, Nebraska City, NE 68410).

For directions and information on constructing homemade birdhouses, write for a free copy of "Recycle for the Birds" to National Wildlife Federation (8925 Leesburg Pike, Vienna, VA 22184).

For data on what can be done about litter, write for a copy of "Deadly Throwaways" to Defenders of Wildlife (1244 19th St. NW, Washington, DC 20036).

If you would like to have your backyard certified as a Backyard Wildlife Habitat, write to the National Wildlife Federation (Backyard Wildlife Habitat Program, 1412 16th St. NW, Washington, DC 20036).

You and your child may also wish to select one of two of the following activities to share together.

1. Have your child look through the local newspaper for articles about local conservation efforts. These articles can be cut from the paper and pasted into a "Conservation Scrapbook." Take time periodically to discuss the implications of these local efforts on the preservation of the entire earth. What are some things the family can do to support local efforts at home?

2. You and your child can write a letter to the editor of your local newspaper. Explain some of your concerns about the environment, congratulate local organizations and agencies for their efforts, and make suggestions on how other citizens can become involved in conservation efforts. You might also write a letter to your state or national representative to explain your concerns.

3. If possible, take your child on a "field trip" to a local recycling plant and garbage dump. Talk with your child about the differences between these two places and the long-term implications. What will happen to the garbage dump in 50 years? What will happen to the recycling plant?

Fun Fact of the Week
The smallest drip from a leaky faucet can waste more than 50 gallons of water a day.

Suggested Children's Books
Cook, David. *Environment.* New York: Crown, 1985.
George, Jean C. *One Day in the Tropical Rain Forest.* New York: HarperCollins, 1990.
Huff, Barbara. *Greening the City Streets: The Story of Community Gardens.* New York: Clarion, 1990.
Newton, James. *A Forest is Reborn.* New York: Crowell, 1982.
Van Allsburg, Chris. *Just a Dream.* Boston: Houghton Mifflin, 1990.

Sincerely,

Physical Sciences

Simple Machines

Dear Parents:

We use machines to do much of the work around our homes. While we frequently use very complex machines powered by some outside force like electricity, many machines are very simple in nature. Simple machines actually form the basis for some of the more complex work-saving devices we use every day. A simple machine is defined as a tool with few or no moving parts that makes work easier.

There are six types of simple machines: the **lever**, the **inclined plane**, the **wheel and axle**, the **screw**, the **wedge**, and the **pulley**. Children are often amazed to discover how often they use some or all of these simple machines each day. For example, when they turn the doorknob on a door, they are using a wheel and axle. When they cut a piece of cheese with a knife, they are using a wedge. When they walk up a flight of stairs, they are using a modified form of an inclined plane.

By involving your child in one or two of the following activities, you can help him or her appreciate simple machines more.

1. Have your child divide a sheet of paper into six columns, each titled with the name of a simple machine. Take your child on a "field trip" around the house and have him or her record the various examples of each of the simple machines in use. Here are some to get you started: Lever—seesaw, crowbar; Inclined Plane—ramp, stairs; Wheel and Axle—can opener, bicycle; Screw—jar lid, pencil sharpener; Wedge—knife, ax; Pulley—clothesline, flagpole.

2. A screw has much more holding power than does a nail. You can demonstrate this to your child by pounding a nail part way into a piece of wood. Insert a small wood screw into the same piece of wood. Ask your child to use a claw hammer to remove both the nail and screw. Discuss which one needed more force to remove it from the wood. Discuss with your child situations in which the greater degree of holding power would be important (house construction, bookshelves, etc.). Also, point out that the hammer is also a simple machine—a lever.

3. Children are often amazed to discover that parts of their own bodies are simple machines, too. For example, since teeth cut through food, they are an example of wedges. Arms are levers that can lift objects. Discuss with your child other parts of the human body that are simple machines. In other words, how many simple machines does your child carry around every day? Ask other family members to help with this activity.

Fun Fact of the Week

The United States and the country of Brunei are the only countries in the whole world that use the standard system (verses the metric system) of measurement.

Suggested Children's Books

Ardley, Neil. *Making Things Move.* New York: Watts, 1984.

Weitzman, David. *Windmills, Bridges, and Old Machines: Discovering Our Industrial Past.* New York: Scribner's, 1982.

Zubrowski, Bernie. *Messing Around with Water Pumps and Siphons.* Boston: Little, Brown, 1981.

Zubrowski, Bernie. *Wheels at Work: Building and Experimenting with Models of Machines.* New York: Morrow, 1986.

Sincerely,

Physical Sciences

Heat

Dear Parents:

Heat is a part of our everyday lives. We use it to cook our food, warm our homes, and protect our bodies. We often talk about how hot or cold something is. Whenever people discuss the weather, the conversation always includes remarks about the temperature.

Children, too, have a fascination with heat. Like adults, they often confuse heat and temperature. **Heat** refers to energy in motion. In other words, heat is the total amount of energy in the moving particles of an object or substance. **Temperature** is a measure of how hot something is. Or, temperature can be considered to be a measure of the amount of heat an object has.

Heat moves through solid objects by a process known as **conduction**—the movement of heat from one molecule to the next. Some objects, such as those made from metal, conduct heat better than others, such as those made of wood. Heat also moves through the air and water through a process known as **convection**—when heat rises and sinks in moving currents.

While we often take heat for granted, it's important that students understand what it is and what it does. I encourage you to use one or two of the activities below with your child to help him or her understand the function and use of heat.

1. CAUTION: Do this activity with adult supervision! Place a pan with a long metal handle on the stove. Have your child put on an oven mitt or thick pot holder and hold the bulb end of a thermometer to the end of the handle. After two minutes, note the temperature of the handle. Turn on the stove and have your child note (while you record) the temperature of the handle at 20-second intervals. What can your child surmise about the conduction of heat through the handle of the pan?

2. Have your child select a room in the house that can be closed off from the rest of the house (a bedroom usually works best). Have your child take the temperature of various parts of the room (in the corner near the ceiling, along the floor of one wall, inside a closet). Talk with your child about reasons why the temperature in a single room may vary by several degrees. Share with your child the process of convection in which air tends to move in currents from the top to the bottom and back up to the top of a room again.

3. Have your child put an ice cube on a plate and record the length of time it takes to melt completely. Give your child some sealable containers and challenge him or her to add some materials that will slow down the rate of melting for an ice cube. Materials which could be added to the containers include shredded newspaper, sawdust, cotton, quilted fabric, leather, wool, popped popcorn, uncooked rice, modeling clay, or sand. Which material slows down the "melting rate" the most? Help your child make a graph comparing the times.

Fun Fact of the Week

It takes five times more heat to turn boiling water into steam than it does to bring freezing water to a boil.

Suggested Children's Books

Ardley, Neil. *Hot and Cold.* New York: Watts, 1985.
Ontario Science Centre. *Scienceworks.* Reading, MA: Addison-Wesley, 1986.
Santrey, Laurence. *Heat.* Mahwah, NJ: Troll, 1985.
Wilkes, Angela and David Mostyn. *Simple Science.* London: Usborne, 1983.

Sincerely,

Physical Sciences

Electricity

Dear Parents:

We often take electricity for granted. We flip light switches and the lights come on. We put toast in toasters and it gets browned. We put food in microwave ovens and it gets cooked. All of these events are possible because of electricity, yet few people understand some of the processes involved when electricity flows.

Electricity is created when tiny bits of matter—known as **electrons**—flow from one place to another. We are familiar with the **current electricity** we use in our homes, but there is also **static electricity**—the type that causes articles of clothing to stick together in the dryer. While current electricity can be controlled, static electricity cannot, as you know whenever you walk across a thick rug and touch a metal doorknob. It's important for students to understand not only the types of electricity, but also the ways in which we can use it and control it.

Of course, in any discussion of electricity it's also important to understand and know some of its dangers. Any and all activities done with electricity should be carefully supervised at home, just as they are at school. One or two of the following activities may help your child appreciate more about electricity.

1. Take a piece of newspaper and tear it into very small pieces. Spread the pieces on a sheet of colored paper and sprinkle some salt and pepper on the paper, too. Rub a plastic comb with a wool cloth several times. Move the comb near the paper pieces, salt, and pepper. Discuss with your child what he or she observes (static electricity).

2. Have your child make a large chart divided into three sections—"Heat," "Light," and "Motion." Ask your child to go through the house and record the names of electrical appliances and devices in their proper categories. Which category has the most names? Why?

3. Be sure to talk with your child about some of the dangers of electricity. You and your child can write and illustrate a family safety guide on the uses of electricity in your home. You can contact your local electric company, electrical contractor, or professional electrician for free materials, ideas, and suggestions.

Fun Fact of the Week

More than 10 percent of the energy sent through power line wires is lost as heat.

Suggested Children's Books

Adley, Neil. *Discovering Electricity.* New York: Watts, 1984.

Math, Irwin. *Wires and Watts: Understanding and Using Electricity.* New York: Scribner's, 1981.

Sootin, Harry. *Experiments with Electric Currents.* New York: Norton, 1969.

Wade, Harlan. *A Book About Electricity.* Milwaukee, WI: Raintree, 1977.

Sincerely,

Physical Sciences

Magnetism

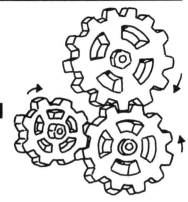

Dear Parents:

One of the topics most frequently discussed and presented in a science program is magnetism. That is because magnets are so much a part of our everyday lives. They are in the motors we use, the toys we play with, the household tools we depend on, and, yes, even on the doors and sides of our refrigerators. Magnets and magnetism are some of the most fascinating areas for study by science students.

A **magnet** is anything that pulls iron and steel to it. **Magnetism** is the force around a magnet. Basically, magnets are of two types—natural and artificial. Natural magnets, or **lodestones**, are a type of iron ore known as **magnetite**. They occur naturally and have north and south poles as do the magnets we use in our homes and schools. Artificial magnets are usually made of steel or alnico (an alloy of aluminum, nickel, copper, and cobalt). These types of magnets are commonly in the shape of a bar or horseshoe. Most of the magnets we use on our refrigerator doors and in our toys are alnico magnets.

Kids enjoy magnets because of the "power" they can use in various games and other activities. While children often think that magnets are toys, they are surprised to discover the wide and varied uses for magnets. In fact, much of the construction work in your town or community would not be possible without the use of magnets. You can help your child learn more about the properties of magnets by selecting one or two of the activities below.

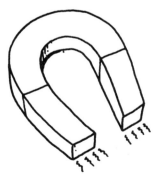

1. **CAUTION: An adult should supervise this experiment. Take special care that children keep their hands away from mouths and eyes. Use safety goggles with younger children.** Obtain a sheet of thin cardboard and some iron filings (available at toy, hobby, or hardware stores). Draw a simple outline of a person's face on the cardboard (head, eyes, nose, mouth, ears). Place four water glasses on a table and place the cardboard so the corners rest on the edges of the glasses. Sprinkle some iron filings in the middle of the illustration. Give your child a strong bar magnet (available at most larger toy stores) and have him or her move it underneath the cardboard so that the iron filings can be moved around the "face." Have your child put a beard, mustache, eyebrows, and other "hair" on various locations on the face.

2. You and your child can create your own magnetic compass. Rub a bar magnet lengthwise on a sewing needle (in one direction only). Cut a small piece of sponge, about one inch square.

Carefully stick the needle through a small portion of the top of the sponge. Mix a few drops of liquid detergent into a small bowl of water. Place the sponge and needle in the bowl. The needle should be parallel to the water, but not touching it. The sponge and needle should float in the center of the bowl. If they don't, add one or more drops of detergent until they do. After the needle stops moving, it is pointing to the earth's magnetic north pole.

3. At a yard or garage sale, obtain one or two small motors. With your child, disassemble the motors and locate the magnets inside each. Explain to your child that most motors have similar magnets in them which are used in the generation of electricity.

Fun Fact of the Week

The largest magnet in the world is the entire planet earth.

Suggested Children's Books

Adler, David. *Amazing Magnets.* Mahwah, NJ: Troll, 1983.
Challand, Helen. *Experiments with Magnets.* Chicago: Childrens Press, 1986.
Kent, Amanda. *Physics.* New York: EDC Publishing, 1984.
Whyman, Kathryn. *Electricity and Magnetism.* New York: Gloucester Press, 1985.

Sincerely,

Physical Sciences

Work and Energy

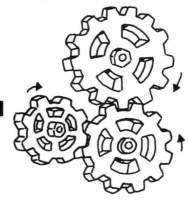

Dear Parents:

When we think of work, we usually think about our jobs, or some of the chores we do around the house. Yet work, to a scientist, is a much more abstract idea than that. As you pick up this letter, you are doing work. You're working as you help your child participate in some of the suggested activities, and as you put this letter away and move on to other activities.

In scientific terms **work** is done only when a force (a push or a pull) makes something move. When we rake leaves, work is done. When we climb a mountain, work is done. However, if you were to push against a solid brick wall, no work would be done, because nothing was moved. (At least, the only work being done would be in your moving muscles.)

Friction is a force that slows down or stops moving objects. Friction results when two objects are in contact with one another. Friction can result from an object moving through the air or moving across another object. When you throw a baseball, friction is one of the reasons why the ball slows down and eventually falls to the ground. The ball is moving against the friction of the wind.

Scientists define **energy** as the ability to do work. A parent who has tried to get a child to clean up his or her messy room might say that the child has not used enough energy. In part, that statement is absolutely correct. Humans get their energy from the food they eat as cars get their energy from gasoline and a hair dryer gets its energy from electricity. It is because of those different forms of energy that each person or machine is able to do work.

We often speak of two types of energy—**kinetic energy**, which is the energy of motion (a ball moving through the air) and **potential energy**, which is stored energy. A ball lying on the ground has potential energy, which is released when you throw it. All objects can have both potential and kinetic energy.

You can help your child understand more about work and energy by selecting one or two of the following activities. Please plan to share and discuss these activities with other members of the family, too.

1. Obtain several different objects of different weights (a book, a small brick, a cooking utensil—each should be between one ounce to one pound in weight). Fasten a rubber band to a piece of string, and tie the string around one object. Repeat this procedure for each of the other objects. Have your child lift one object by the rubber band. When the band is stretched to its fullest, measure its length with a ruler. Use the same (or an identical rubber band) for each of the other objects (measuring the length of each when a different object is lifted). Explain that the longer the rubber band becomes, the greater the amount of energy needed to lift an object. Explain that heavier objects, therefore, require more energy to lift.

2. You can demonstrate friction by having your child rub his or her hands together rapidly. The heat that results occurs because of the friction between the two hands. Have your child move pairs of other objects together. Move a wood block across a brick several times. Rub a coffee cup across a carpet, or a shoe across concrete. Have your child touch one of the two objects and note the amount of heat generated. That heat is the result of friction between the objects.

3. Challenge your child to make a list of examples when energy is used, but work is not done. For example, a hair dryer that does not generate heat; you pushing against the side of the house; or you trying to open a jar lid without success. Talk with your child about some of the different kinds of work that family members do around the house each day. Ask your child to determine if the amount of work done is equal to the amount of energy used. (Most often, there will usually be more energy used than work accomplished.)

Fun Fact of the Week
California produces about 90 percent of the wind-generated energy in the United States.

Suggested Children's Books
Ardley, Neil. *Making Things Move.* New York: Watts, 1984.
Chase, Sara. *Moving to Win: The Physics of Sports.* New York: Messner, 1977.
Zubrowski, Bernard. *Messing Around with Water Pumps and Siphons.* Boston: Little, Brown, 1981.
Zubrowski, Bernard. *Raceways: Having Fun with Balls and Tracks.* New York: Morrow, 1985.

Sincerely,

Physical Sciences

Light

Dear Parents:

While we may take light for granted, without it life as we know it would not be possible. Most plants would not be able to grow. People would not be able to see the world around them. We would not have any food to eat.

Turn on a flashlight and you see light. Flip a switch and you light your home. Look outside at night and see lights everywhere. By definition, light is a visible form of energy. White light, the most common form of light, is actually a combination of all the colors of the rainbow (red, orange, yellow, green, blue, indigo, and violet). Both white light and the colors of light in it are referred to as the **visible spectrum**.

People see colors because of what happens when light hits different objects. Objects **absorb**, or take in, some of the light that hits them. Objects also **reflect**, or bounce back, some light. An opaque object (one in which light cannot pass through) reflects whatever light it does not absorb. The color of the object is actually the color of the light it reflects. For example, a green shirt absorbs all the colors of the spectrum except green. Green is the color it reflects, and hence the color we see. A white object reflects all colors.

Children are intrigued by the properties of light. When they use mirrors, they see how light reflects off flat surfaces. You and your child may wish to discover some of the other interesting aspects of light by selecting one or two of the activities below.

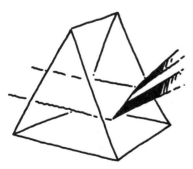

1. Visit a local toy store or hobby store and obtain a glass prism. Have your child use the prism to create rainbows throughout the house. By holding the prism in a beam of white light, you see the colors of the rainbow separated. You can then show the colors on a piece of paper or on a nearby wall. (You can also obtain a very inexpensive prism [catalog No. 57-160-4679] from Delta Education, P.O. Box 950, Hudson, NH 03051).

2. Have your child make a chart with the words "Transparent," "Translucent," and "Opaque" listed across the top. (Transparent objects are those that light passes through readily, such as glass and clear plastic. Translucent objects are those through which some light passes, but through which you cannot see clearly, such as waxed paper and shower doors. Opaque objects, such as wood or steel, permit no light to pass through.) Have your child take a "field trip" through the house and make a list of objects that could be placed into one of the three categories. Which category has the fewest number of items?

3. Mirrors have smooth, shiny surfaces that form good images. Discuss with your child the various ways in which people use mirrors in everyday life. Discuss mirrors in bathrooms, on and in cars, in department stores, and at the dentist's office. Challenge your child to develop a list of as many different uses for mirrors as possible. (How do scientists use mirrors to watch the stars? To see tiny objects?) What are the benefits of this accurate reflection of light?

Fun Fact of the Week

It would take a beam of light 100,000 years to cross from one side of the Milky Way Galaxy to the other.

Suggested Children's Books

Branley, Franklyn. *Color: From Rainbows to Lasers.* New York: Crowell, 1978.
Laurence, Clifford. *The Laser Book.* New York: Prentice-Hall, 1986.
Schneider, Herman and Nina. *Science Fun with a Flashlight.* New York: McGraw-Hill, 1975.
Simon, Seymour. *Mirror Magic.* New York: Lothrop, 1980.

Sincerely,

Physical Sciences

Sound

Dear Parents:

No matter where we go or what we do, sound is all around us. From the time we wake up in the morning until we retire at night we are surrounded by all kinds of sounds—loud and soft sounds, pleasurable and unpleasurable sounds, and machine and human sounds.

Although we are constantly surrounded by sounds, many people would find it difficult to describe what sound is. **Sound** is produced when matter vibrates. When a musician strums a guitar, the vibrating strings create sounds. When people talk, their vocal cords vibrate and create sound. When phones ring, pieces of metal vibrate so that sound is produced. Sounds travel through the air in waves, much like the waves in the ocean.

Sound has several different qualities. **Volume** is the loudness or softness of a sound. The loudness of a sound is measured in units known as **decibels**. (The human ear can hear decibels of 0-85 without sustaining any damage.) **Pitch** describes how high or low a sound is (a flute vs. a bass drum, for example). **Frequency** is the speed at which an object vibrates to create sound.

You can help your child learn more about sound by selecting one or two of the following activities. You and your child are encouraged to create additional activities to demonstrate the properties of sound.

1. Obtain several books of different sizes. Have your child put a rubber band around each book so that the rubber bands are each stretched to different lengths. Have your child strum each band to determine which ones produce high-pitched sounds and which produce low-pitched sounds. Does a long rubber band create a high or low sound? How about a short rubber band?

2. If possible, borrow a tuning fork from a music teacher or from a music store. Tap the tuning fork and immediately place the fork tips barely into a bowl of water. Have your child observe and comment on what happens to the water. Have your child note that the vibrating fork creates sound and that when the fork is placed in the water it creates waves similar to the sound waves that travel through the air.

3. Take your child to a concert, opera, symphony, or other musical performance. Talk with your child about the different musical instruments used. Upon your return home, have your child look through some old magazines for pictures of some of the instruments. Ask your child to create two separate posters of those pictures—one of instruments with high pitch and one of instruments with low pitch.

Fun Fact of the Week
A bat will emit sounds at frequencies of up to 230,000 vibrations per second.

Suggested Children's Books
Ardley, Neil. *Music and Sound.* New York: Watts, 1985.
Kettlecamp, Larry. *The Magic of Sound.* New York: Morrow, 1982.
Knight, David. *All About Sound.* Mahwah, NJ: Troll, 1983.
Kramer, Anthony. *The Magic of Sound.* New York: Morrow, 1982.

Sincerely,

Earth Sciences

Seasons

Dear Parents:

New plants begin to poke their heads through the soil, the days are filled with lots of sunshine. Leaves begin to turn brown and scatter with the wind. A blanket of snow covers the ground. These are the signs of the four seasons—spring, summer, autumn, and winter. To people in the north and northeast portions of the United States, these are familiar; for people in the south and southwestern portions of the country, some of these seasonal signs may be unfamiliar. But no matter where you live, seasonal changes affect your life.

Seasons are the result of the earth's orbit around the sun. The earth's **orbit** is the path it takes as it revolves around the sun. One full orbit of the earth around the sun is known as a **revolution**. The earth takes 365 days, or one year, to make one revolution.

Because the earth is tilted on its **axis** (an imaginary line that passes from the North Pole through the earth to the South Pole), sunlight hits the earth at different angles. When the northern part of the earth is tilted toward the sun, it gets more sunlight and experiences summer. At the same time, the southern part of the earth is tilted away from the sun. It receives less sunlight and experiences winter.

In the Northern Hemisphere (the part of the earth in which we live) the seasons usually begin on or near the following dates: Spring—March 20; Summer—June 21; Fall—September 22; and Winter—December 21. In the Southern Hemisphere (Australia, Argentina, South Africa, etc.), the seasons are "reversed." While we are experiencing summer, people in Australia are having winter; when we have winter, people in South Africa are having summer.

You can help your child learn more about the seasons by selecting one or two of the activities that follow. Be sure to invite other family members to participate whenever possible.

1. Have your child make a chart that lists the names of the four seasons across the top. Write the words "Plants," "Animals," and "Humans" down the left side of the chart. Ask your child to do some library research and record some of the activities of plants, animals, and humans during each of those seasons. (For example, some animals hibernate during winter, some plants flower in the spring, some humans rake leaves in the fall.) Are there activities that can be listed in more than one column? Are there activities that belong in more than one row?

2. You and your child might want to assemble a collection of newspaper articles about the seasons into a three-ring binder or notebook. This would be an excellent year-long activity. Work with your child to select articles in your daily newspaper that describe an approaching season, feature some of the weather patterns of a particular season, discuss clothing appropriate for a new season, or other pertinent pieces. Enter each article in the appropriate section in the notebook. Plan time to discuss the changes and differences in seasons as depicted in the collection of articles.

3. Obtain a large Styrofoam ball from a hobby store. Push a pencil through the center of the ball and draw a line around the middle of the ball. Take the shade off a table lamp and hold the ball so that the top of the pencil (the North Pole) is tilted slightly toward the light. (The ball should be about one foot away from the bulb). Move the ball in a circle around the light (keeping the angle of the tilt the same). Have your child note when the top half of the ball is closest to the light (summer) and when the top half is farthest from the light (winter). Discuss how this is similar to the movement of the earth and its seasons. Tip: For a better sense of relative sizes, use a large globe lamp or lantern flashlight for the sun. Use a table tennis ball for the earth.

Fun Fact of the Week
The mountains in California get more snow in the winter than the North Pole does all year long.

Suggested Children's Books
Branley, Franklyn. *Sunshine Makes the Seasons*. New York: Crowell, 1985.
Goennel, Heidi. *Seasons*. Boston: Little, Brown, 1986.
Provensen, Alice. *A Book of Seasons*. New York: Random House, 1976.
Yolen, Jane. *Ring of Earth: A Child's Book of Seasons*. New York: Harcourt, 1986.

Sincerely,

Earth Sciences

Water

Dear Parents:

Without question, water is one of the most precious commodities on earth. It is necessary for our health and survival, important in business and industry, and vital for the sustenance of life in all its forms.

Children need to recognize the importance of water in their everyday lives, too. While we frequently turn on a water faucet without really thinking about the liquid that pours out of that tap, students should know that many complex processes take place in order for that water to be available whenever needed.

Obviously, one of the major concerns we have about water today is its purity. Pollution of our water sources and resources is an increasing social, economic, and environmental issue. Helping children appreciate the importance of water in their lives can help them understand their roles in preserving this valuable resource.

Select one or two of the following activities to share with your child. Explain that water will be an important concern throughout his or her life.

1. Ask your child to make a list of all the different ways in which water is used in your home (cooking, drinking, bathing). Ask your child to try to compute the average amount of water used in the home each day. If you use public water, you might be able to use your monthly water bill to compute a daily average.

2. You and your child can create your own well. Obtain a large (#10) aluminum can (a coffee can works well). Place a cardboard tube (from a roll of paper towels) upright in the can. Pour a layer of gravel inside the can around the outside of the tube. Pour another layer of sand on top of the gravel. Pour water on the sand until the water reaches the top of the layer of sand. Have your child notice what happens inside the tube. Explain that this is the same process used for obtaining well water.

3. Obtain several different water samples from in and around the house (such as tap water, water from a standing puddle, or rain water). Place coffee filters over several glass jars. Have your child pour each of the water samples into a separate jar. Ask your child to note the impurities that have been "trapped" by each of the filters. Which water has the most impurities? Which water looks cleanest?

Fun Fact of the Week

99.5 percent of all the fresh water on the earth's surface is frozen in glaciers and icecaps.

Suggested Children's Books

Arnold, Caroline. *Bodies of Water: Fun, Facts, and Activities.* New York: Watts, 1985.
Gardner, Robert. *Water: The Life Sustaining Resource.* New York: Messner, 1982.
Hanmer, Trudy. *Water Resources.* New York: Watts, 1985.
Nixon, Hershell and Joan Nixon. *Glaciers: Nature's Frozen Rivers.* New York: Dodd, 1980.
Robin, Gordon. *Glaciers and Ice Sheets.* San Diego, CA: Harcourt, 1984.

Sincerely,

Earth Sciences

Rocks and Soil

Dear Parents:

If your child has ever tracked mud across your clean kitchen floor, you know a lot about rocks and soil. If there's one fact of life that truly stands the test of time, it's that kids and dirt are soon attracted to each other.

There are actually three different kinds of rocks throughout the world. **Igneous** rocks are those that form from melted minerals. They can often be found near volcanoes. Granite is one type of igneous rock. **Sedimentary** rocks are those that are usually formed under water as a result of layers of sediment pressing down on other layers of sediment. Sandstone and limestone are examples of sedimentary rocks. **Metamorphic** rocks are those that result from great heat and pressure inside the earth's surface. Marble is an example of a metamorphic rock.

Soil is actually rocks that have been broken up into very fine pieces. Soil is usually created over several years (hundreds or thousands) and is the result of weathering, erosion, and freezing. The climate and slope of the land also affect how fast soil forms in a particular area. Soil also contains air, water, and decayed matter (known as **humus**). Basically, there are three types of soil—clay, sandy, and loam (a rich mixture of clay, sand, and humus).

Help your child learn more about rocks and soil by selecting one or two of the following activities. Whenever possible, plan to include other family members in your activities.

1. Provide your child with several sealable plastic sandwich bags. Take a "field trip" through your town or neighborhood and invite your child to collect as many different soil samples as

possible. Upon your return home, have your child gently pour each sample onto a white sheet of paper. Give your child some toothpicks and a hand lens (available at most toy or hobby stores). Ask your child to carefully sift through each sample to determine its components. What "ingredients" does your child find in each sample? Are the samples distinctively different or about the same? How big or small are the particles in each sample? Which sample would be best for growing plants?

2. Visit a local garden center or nursery with your child. Take some time to talk with one of the workers about soil conditions in your area. What recommendations would that person make for turning the native soil into the best possible growing medium? What special nutrients or additives should you add to the soil in order to begin a garden? What is distinctive about the native soil that makes it appropriate or inappropriate for growing vegetables, for example?

3. Obtain some organic clay and some modeling clay from a local arts and crafts store. Have your child note the difference in the composition of the two clays. Have your child make a simple piece of pottery from each sample. Place each piece of pottery in the sun. After a few days, ask your child to note the difference between the two pottery pieces. What has changed? What has remained the same? You might wish to explain to your child that pottery pieces more than 5,000 years old have been found at various archeological sites around the world.

Fun Fact of the Week
Every year in the United States seven billion tons of topsoil are washed away.

Suggested Children's Books
Fichter, George. *Rocks and Minerals.* New York: Random House, 1982.
McGovern, Tom. *Album of Rocks and Minerals.* Chicago: Rand McNally, 1981.
Rinkoff, Barbara. *Guess What Rocks Do?.* New York: Lothrop, 1975.
Selsam, Millicent. *First Look at Rocks.* New York: Walker, 1984.

Sincerely,

Earth Sciences

Changes in the Earth: Volcanoes and Earthquakes

Dear Parents:

Visit any science fair in the country and what you're likely to see is an abundance of volcano models and earthquake demonstrations. Unquestionably, kids are fascinated by these major, earth-shaking events. The violence and destruction volcanoes and earthquakes cause are always newsworthy.

Volcanoes occur when melted rock (known as **magma**) is squeezed up through the surface of the earth. Volcanoes occur all over the face of the earth, but most of them are scattered around the edge of the Pacific Ocean in a line known as the "Rim of Fire."

An earthquake, on the other hand, is a shaking or sliding of the earth's crust along fault lines. Earthquakes can be minor or very dangerous. Some of the most severe earthquakes in the United States occur along the San Andreas Fault, which runs along the western coast of California.

You can help your child appreciate the power and magnitude of earthquakes and volcanoes with one or two of the following activities. No matter where you live in the country, your child can learn a great deal about these two powerful forces of nature.

1. Obtain two different colors of modeling clay from a nearby hobby store. Flatten each into a square about 4" on each side. Place the squares on a smooth surface side by side. Ask your child to place his or her hands on the outer edges of the squares and attempt to push them together. Note what happens to the clay as it is slowly pushed together. Inform your child that this demonstrates what happens to the earth's surface during an earthquake.

2. CAUTION: This activity should be done under adult supervision. You and your child can create your own model of a volcano. Make a model of a volcano using modeling clay. Make a depression in the top of the "mountain" and place an empty film can or doll's dish in the depression. Place 1 to 2 tablespoons of baking soda in the container. Slowly and carefully pour in about 1/2 ounce of vinegar into the container. Discuss with your child the reaction that takes place. How is this similar to the action of volcanoes? (Gases build up inside the container and explode through the only opening.)

3. With your child, start a scrapbook of newspaper articles about earthquakes and volcanoes around the world. Clip articles and photos from your daily newspaper and arrange them into a three-ring binder. Have your child find the locations of these events on a large globe or map of the world. Over the course of several months, where do most of these events take place?

Fun Fact of the Week

Millions of earthquakes occur every year. However, only a small fraction—about 3,000—noticeably move the earth.

Suggested Children's Books

Foder, R. V. *Earth Afire! Volcanoes and Their Activity.* New York: Morrow, 1981.

Lauber, Patricia. *Volcano: The Eruption and Healing of Mount St. Helens.* New York: Bradbury, 1986.

Navarra, John. *Earthquake!* New York: Doubleday, 1980.

Nixon, Hershell and Joan Nixon. *Earthquakes: Nature in Motion.* New York: Dodd, 1981.

Simon, Seymour. *Volcanoes.* New York: Morrow, 1981.

Sincerely,

Earth Sciences

Changes in the Earth: Weathering and Erosion

Dear Parents,

The earth is constantly changing. Most of the changes that take place on the earth are natural—earthquakes, erosion, floods. These natural occurrences change the face of the earth—sometimes violently, sometimes slowly over a long period of time.

Change is a natural and normal feature of the planet we live on. While humans have certainly had a profound effect on the shape of the earth, nature has played an even larger role in molding the earth into its various patterns and shapes. The forces of **weathering** (the breaking and wearing away of rocks) and **erosion** (the moving of weathered rocks, and soil by wind, water, or ice), for example, are some of the most powerful forces in nature. The Grand Canyon in Arizona, Mammoth Caves in Kentucky, and the Mississippi River delta are examples of weathering and erosion at work.

You can help your child understand the impact of erosion and weathering with one or two of the following activities. Most important, however, is the need to help your child understand his or her role in preserving the earth through conservation efforts and practices.

1. Obtain some sandstone rocks from your area or from a local building supplier or home improvement center. Soak several rocks in water overnight. Place the soaked rocks in one plastic sealable bag and the unsoaked rocks in another bag. Put both bags in the freezer and leave for 24 hours. Remove the bags from the freezer and talk with your child about the reasons why the soaked rocks broke apart (the water froze and expanded, cracking the rocks). Discuss the relationship between this demonstration and natural occurrences that take place in nature over a long period of time (years and decades).

2. Have your child mix together several tablespoons of white glue and sand into a stiff mixture. Form the mixture into a cube and allow to dry for two days. Place the cube in a plastic margarine tub with some water and several small pebbles. Have your child shake the tub for five minutes. Take out the cube and discuss reasons why it is smaller. How would the action of water and rocks affect the shape of a canyon, for example, over a period of thousands of years? You might want to show your child some pictures of the Grand Canyon to illustrate the effects of erosion over time.

3. You can demonstrate erosion to your child through the use of everyday objects. For example, coins that have been worn smooth, old shoes with the heels worn down, or an old bicycle tire with no tread. Discuss reasons why these objects (or any other objects) tend to wear down over time.

Fun Fact of the Week
Each year the world's deserts increase by as much as 16,000 square miles.

Suggested Children's Books
Pringle, Lawrence. *Restoring Our Earth.* New York: Enslow, 1987.
Stille, Darlene. *Soil Erosion and Pollution.* Chicago: Childrens Press, 1990.
Wheeler, Jill. *The Land We Live On.* Minneapolis: Abdo and Daughters, 1990.
Wyler, Rose. *Science Fun with Mud and Dirt.* New York: Simon and Schuster, 1986.

Sincerely,

Earth Sciences

Oceans

Dear Parents:

Covering nearly 70 percent of the earth's surface, the oceans of the world are vital to the planet. The four major oceans—the Pacific, Atlantic, Indian, and Arctic—are all really part of one large world ocean.

Students need to understand that, even though they may not live near an ocean, oceans have a great deal of importance in their lives. Most of the water from rain, and all of the water in the world's rivers, eventually reaches the ocean. People obtain many resources from the ocean. About 30 percent of the world's salt comes from the oceans. Fish and other types of seafood, oil, and gas are other ocean resources. So, no matter where your family lives, it probably has used some of the ocean's resources at least once or twice in the past week.

You can help your child appreciate the benefits and properties of the ocean with one or two of the following activities. Help your child understand that many of the things we use everyday are possible because of ocean resources.

1. Mix 1 cup of salt with 2-1/2 cups of water. Pour the mixture into an aluminum pie plate. Pour 2-1/2 cups of regular tap water into another pie plate. Place both plates in the sunlight and leave them outside for several days. Discuss with your child what you discover in each pan.

2. Look through several old magazines and locate pictures of animals and plants that live in the ocean. Cut out the pictures and paste them onto a large piece of cardboard to create an attractive collage. You and your child can create an alternate collage displaying pictures of products we get from the ocean.

3. Fill an empty 2-liter soda bottle 1/3 of the way up with salad oil. Fill the rest of the bottle (all the way to the brim) with water dyed with blue food coloring. Put on the top and lay the bottle on its side. Have your child tip the bottle back and forth. How does the movement inside the bottle relate to the movement of waves across the surface of the ocean?

Fun Fact of the Week

The deepest place in the oceans is the Marianas Trench in the Pacific Ocean. The bottom of this trench is about 36,163 feet below the surface of the water.

Suggested Children's Books

Malnig, Anita. *Where the Waves Break: Life at the Edge of the Sea.* Minneapolis: Carolrhoda, 1985.

Morris, Rick. *Mysteries and Marvels of Ocean Life.* New York: Random House, 1981.

Robinson, Howard (Ed.). *Amazing Creatures of the Sea.* Washington, DC: National Wildlife Federation, 1987.

Williams, Brian. *Secrets of the Sea.* New York: Ray Rourke, 1981.

Sincerely,

Earth Sciences

Weather

Dear Parents,

Mark Twain used to say, "Everybody talks about the weather, but nobody does anything about it!" It's certainly true that weather affects our daily lives—both in work and play. Many of our activities both in and out of the home are influenced to a significant degree by the weather.

Children are aware of the effects of weather from a very early age. Often weather is their first contact with the world of science. The games they play, the clothes they wear, and the places they travel are influenced by weather and its patterns.

You can help your child appreciate weather and how it is measured with some of the following activities. It will be important to discuss the changes that can occur in the weather and how those changes affect many of the things you and your child do.

1. Work with your child to collect a week's worth of weather forecasts from the daily newspaper. Cut out each of these and paste them vertically down the left side of a long sheet of paper (or several pieces of paper taped together). For each weather prediction have your child write the actual weather for the designated day in a column down the right side of the paper. Talk with your child about any differences between the forecasts and the actual weather. How can your child account for any differences?

2. If possible, obtain an inexpensive weather station that can be set up at home. Many hobby and toy stores carry home weather stations. Or you can order some weather instruments (anemometer, thermometers, rain gauge, humidity detector, etc.) directly from Delta Education, P.O. Box 950, Hudson, NH 03051 (800-442-5444).

3. Show your child examples of newspaper weather maps (*USA Today* has particularly colorful and detailed maps in its daily editions). Explain to your child some of the symbols used on the map and what they mean for your area of the country.

Fun Fact of the Week
The surface of the earth is struck by approximately one hundred bolts of lightning every single second.

Suggested Children's Books

Craig, Jean. *Questions and Answers About Weather*. New York: Four Winds, 1969.

Gibbons, Gail. *Weather Forecasting*. New York: Four Winds, 1987.

Moncure, Jane. *What Causes It?* Chicago: Childrens Press, 1977.

Webster, Vera. *Weather Experiments*. Chicago: Childrens Press, 1982.

Sincerely,

Earth Sciences

Weather and Climate

Dear Parents:

If you live in southern California, you know what kind of weather you can expect throughout the year. If you live in northern Maine you also know what kind of weather you can expect. You also know that the year-round weather patterns in Fort Kent, Maine, will be quite different from the weather patterns experienced by people living in San Diego, California.

People often confuse weather and climate. **Weather** refers to changes in temperature and precipitation (rain, snow, sleet, hail) that occur on a day-to-day basis. **Climate,** on the other hand, is an average of weather conditions in an area over a long period of time. For example, if you live in San Diego, the weather today may be cool and foggy, but the annual climate is typically temperate. While weather is often unpredictable, climate is always predictable. Both weather and climate of an area are determined by its location on earth, its proximity to large bodies of water, and its elevation. Although people may complain about the weather (on a day-to-day basis), they typically live in an area because they know what they can expect (in terms of climate) during the course of the year.

You can help your child appreciate and understand the differences (and similarities) that exist between weather and climate with one or two of the activities below. Whenever possible, try to involve several family members in these suggested activities.

1. Look at the weather section of your local newspaper. Have your child put together several weather notebooks of the weather patterns in selected U.S. cities. Have your child select four or five cities to "track" over the course of several weeks or months. It would be valuable to choose cities in different regions. For example, cities near large bodies of water (San Francisco, Baltimore); cities in mountainous regions (Denver, Boise); cities in dry regions (Phoenix, Dallas); and cities in semitropical areas (Miami, New Orleans). Ask your child to record, on a daily basis, the temperatures, rainfall, cloud conditions, humidity and other pertinent data recorded in the newspaper. After some time discuss with your child the predicted climate for each of the selected cities (based on an average of the daily weather conditions over time).

2. You and your child may wish to set up a family weather station and track the weather conditions in your area over an extended period of time (several months). An inexpensive (about $5.00) weather station manufactured by NASCO can be found in many toy stores. You may also wish to order a Daily Weather Log from Edmund Scientific (101 East Gloucester Pike, Barrington, NJ 08007 or 1-800-257-6173).

3. Cities near large bodies of water have constant climates, while cities distant from bodies of water have climates in which the temperatures can vary widely. This is because land absorbs solar energy quickly during the day, but loses heat quickly at night. You can help your child understand this concept by filling two cups of equal size—one with soil or sand, the other with water. Place a thermometer in each and put the cups in the sun for 30 minutes. Ask your child to record the temperatures. Then place the cups in the refrigerator for five minutes and read and record the temperatures again (your child should discover that the soil absorbed and lost heat more quickly than did the water).

Fun Fact of the Week

The lowest temperature ever recorded at the South Pole was -126.9^0F. The lowest temperature ever recorded at the North Pole was -96^0F.

Suggested Children's Books

Briggs, Carole. *Research Balloons: Exploring Hidden Worlds.* Minneapolis: Lerner, 1987.
Compton, Grant. *What Does a Meteorologist Do?* New York: Dodd, Mead, 1981.
Gay, Kathryn. *Acid Rain.* New York: Watts, 1983.
Pringle, Lawrence. *Global Warming.* New York: Arcade, 1989.

Sincerely,

Earth Sciences

Clouds and Storms

Dear Parents:

Watch any local TV weather report and there will usually be some mention of clouds or storms. Terms such as "partly cloudy," "partly sunny with a likelihood of rain," or "thunderstorms this afternoon" are a common part of many weather forecasts. We often plan vacations, picnics, and other social events around the degree of cloudiness or chance of storms that may occur on a specific day or weekend.

Clouds are formed when warm air rises and cools in the upper atmosphere. As the air cools, the water vapor condenses into tiny droplets of water, which come together to form clouds. Usually, these droplets condense around a microscopic particle of dust, smoke, or salt in the air. This is what gives clouds their color. Clouds are usually of three types—**cirrus** or feathery clouds (usually indicative of good weather), **cumulus** or cottony clouds (usually associated with good weather, but can become stormy when large); and **stratus** or layered clouds, which are frequently associated with rainstorms.

Three of the major types of storms are thunderstorms, hurricanes, and tornadoes. **Thunderstorms** occur when warm moist air rises quickly in the sky. As cool air replaces the rising air, winds begin to blow. Large clouds begin to form and rain eventually begins to fall. **Hurricanes** are large storms that form over the ocean. Warm air rises from the ocean surface and is replaced by cool air. This causes a spinning motion that creates powerful winds. The destruction Hurricane Andrew caused in Miami and New Orleans in August of 1992 is an example of this power. A **tornado** forms during a thunderstorm when air rises very quickly. The air twists into a funnel-shaped cloud that moves along a narrow path. The wind speed of a tornado may reach 250 MPH.

You can help your child learn more about clouds and storms by selecting one or two of the activities below. Plan some time in which your child can share his or her discoveries with other members of the family.

1. CAUTION: An adult should supervise this activity. It's best done outdoors or in a sink or tub. You and your child can create a miniature tornado. Obtain two 2-liter soda bottles. Fill one bottle two-thirds full of water. Place the other bottle upside down on top of the water-filled bottle and tape the two bottle tops tightly together. Invert the bottles and, holding both bottles firmly, spin the empty bottle (the one on the bottom) around on its rim in a clockwise motion. The water inside the top bottle will begin to spin clockwise as it descends into the bottom bottle.

The shape of that spinning will be similar to a tornado on land. (Note: You may have to practice with the two bottles several times to duplicate the proper motion.)

2. Your child might keep a cloud diary. Record the types of clouds that appear in your area over the space of one or two weeks. Next to each cloud type, have your child record the type of weather associated with that cloud. Have your child compare his or her notes with information in the weather section of your local newspaper. How accurate are clouds in identifying the type of weather experienced during a specific day? How accurate are clouds in predicting the weather for the next day?

3. During the summer and fall months, when most hurricanes and tornadoes occur, have your child maintain a "Storm Journal"—a record of the storms that occur at various places throughout the United States. Have your child cut out newspaper articles from the local newspaper and arrange them into a three-ring binder. The newspaper *USA Today* has very colorful weather maps and charts in each daily edition. Several of these can also be included in the journal.

Fun Fact of the Week

Each year, there are an estimated 16 million thunderstorms throughout the world.

Suggested Children's Books

Alth, Max and Charlotte. *Disastrous Hurricanes and Tornadoes.* New York: Watts, 1981.
Branley, Franklyn. *Flash, Crash, Rumble and Roll.* New York: Crowell, 1985.
Ford, Adam. *Weather Watch.* New York: Lothrop, 1982.
Simon, Seymour. *Storms.* New York: Morrow, 1989.

Sincerely,

Space Sciences

The Earth

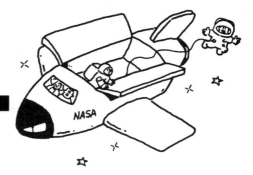

Dear Parents:

Four and a half billion—it's a number almost too large to comprehend. Yet that's how many years the Earth has been in existence. During that time the planet has undergone some remarkable changes. Rocks have formed, primeval seas have ebbed and flowed across vast continents, and dramatic weather conditions have contributed to the geography and structure of our planet. Still, it's amazing to realize that this planet is only a microcosm in the vastness of the universe. It is but one particle in a universe of stars, satellites, meteorites, and other celestial bodies. The beauty of our world and its place in the universe are areas ripe for exploration.

Knowledge of our world contributes not only to an appreciation of its existence, but also to an initiative to preserve it. Thus, it is vitally important for students to get a sense of the majesty of the planet on which they live and the overwhelming need to preserve its life forms.

While any study of the space sciences should begin with a study and understanding of the planet Earth, children should also know that the earth is one part of a very complex solar system, the Milky Way Galaxy, and the universe. In short, the planet Earth, is just one small part of the enormity of space. Comprehending our planet will help youngsters comprehend some of the other parts of space, too. You can help your child learn more about our planet by selecting one or two of the following activities.

1. You and your child might enjoy receiving some aerial photographs of Earth taken from satellites, space shuttles, and other space craft. Write to the Earth Resources Observation Systems Data Center (U.S. Geological Survey, Sioux Falls, SD 57198) and request information on the availability of specific photos.

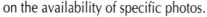

2. If your child is interested in maps and mapping, write to the American Association for the Advancement of Science (1776 Massachusetts Avenue, NW, Washington, DC 20036) and ask for the "Maps and Mapping" pamphlet. The pamphlet is part of the Association's *Opportunities in Space* series.

3. Obtain a medium-size Styrofoam ball from a local hobby or art store. Poke a pencil through the ball. Have your child trace a large circle on paper with the pencil. Explain to him or her that the ball is moving like the earth revolving around the sun. Then, have your child spin the ball on the pencil. This represents the earth rotating. Have your child retrace the original circle while spinning the ball. This demonstrates the revolution (the earth spinning around on its axis once every 24 hours) and rotation (the annual spinning of the earth around the sun).

Fun Fact of the Week

The average orbital speed of the earth around the sun is 66,641 MPH.

Suggested Children's Books

Coburn, Doris. *A Spit is a Piece of Land: Landforms in the U.S.A.* New York: Messner, 1978.
Ford, Adam. *Spaceship Earth.* New York: Lothrop, 1981.
Gallant, Roy. *Our Restless Earth.* New York: Watts, 1986.
Hirst, Robin and Sally. *My Place in Space.* New York: Watts, 1990.

Sincerely,

Space Sciences

The Sun, Moon, and Stars

Dear Parents:

How does the moon stay up in the sky? Where do the stars go during the day? Where does the sun go at night? When children are young, they are full of questions about the sun, moon, and stars. As they get older they begin to develop an awareness of the importance of these celestial bodies to life here on earth. The sun is the source of all life on our planet. The sun provides light and warmth that are necessary for the survival of all living things. People throughout history have looked at the moon with mystery and speculation. The moon's gravity influences tidal patterns on the earth. The stars have fascinated humans for centuries. Many of the stars have been "arranged" into patterns known as constellations.

The sun, moon, and stars, as the celestial bodies most visible in the sky, help students appreciate the largeness of the solar system (the nine planets, their moons, and the sun) as well as the vastness of space. Students can learn that without the sun, life as we know it would not exist. Without the moon, many of our daily activities as well as much of the commerce of the world would be severely curtailed. And without the stars, we would not be able to decipher many of the mysteries of outer space.

You can help your child appreciate the sun, moon, and stars. Plan to share one or two of the following activities with your child.

1. You and your child might enjoy making sun pictures. Obtain several different colored sheets of construction paper from a local stationery or office supply store. Ask your child to draw an illustration of a specific object (an animal, building, person) and cut it out. Lay that illustration on top of a sheet of dark-colored construction paper (you and your child may want to create several of these). Place the sheets out in the sun and leave them there for several days. Afterwards, remove the illustrated sheet from the top of the dark-colored sheet. Note how the sun has "outlined" the figure.

2. The gravity on the moon's surface is one-sixth that of the earth. That is, a person weighing 120 pounds on the earth would only weigh 20 pounds on the moon. Ask your child to compute the "moon weights" of several family members, relatives, and friends. He or she might wish to make a comparative chart of earth weights and moon weights.

3. Obtain a variety of flashlights and give one to each of several different family members. Darken a room and ask your child to stand near the front. Have family members place themselves at different distances away from your child and shine their flashlights at him or her. Point out to your child that when stars are close to the earth they appear large (as do the flashlights); when stars are far away they appear small. Inform your child that scientists can often measure the distance stars are from the earth by the intensity (or brightness) of their light. (Note: This experiment would be especially effective outdoors on a dark night.)

Fun Fact of the Week

Temperatures on the surface of the moon reach a high of 243°F and a low of -261°F.

Suggested Children's Books

Fields, Alice. *The Sun.* New York: Watts, 1980.
Simon, Seymour. *The Moon.* New York: Four Winds, 1984.
Simon, Seymour. *Stars.* New York: Morrow, 1986.
Simon, Seymour. *The Sun.* New York: Morrow, 1986.

Sincerely,

Space Sciences

The Planets

Dear Parents:

Some are large, some are small. Some are nearby, some are far away. Some are very hot, and some are very cold. What are they?—the planets of our solar system. For thousands of years, humans have been fascinated by the planets, those seen as well as those unseen.

Children, too, enjoy discovering facts about the planets. For example, did you know that three of the planets have rings? That a year on Mercury is only 88 days long? That Jupiter has sixteen moons? That the order of the planets will change sometime during the year 1999? Each of the planets has very unique features that distinguish it from the other planets. Those differences are areas ripe for discovery and exploration.

You can help your child learn more about the planets by choosing one or two of the suggested activities below. If possible, you may also wish to take a "family field trip" to a nearby observatory or planetarium to obtain more information on the planets.

1. Provide your child with several sheets of paper. Have him or her cut out nine circles, each representing one of the nine planets. So that your child will be able to understand the relative sizes of the planets in relation to each other, use the following measurements (each measurement represents the diameter of a circle):

Mercury—0.4"; Venus—0.9"; Earth—0.9"; Mars—0.5"; Jupiter—11.2"; Saturn—9.4"; Uranus—3.7"; Neptune—3.8"; and Pluto—0.5".

2. Your child can construct a mobile of the planets in their correct order. Obtain some Styrofoam balls at a local hobby store and paint each a different color. Have your child push a pin into each and tie a length of thread to each pin. The balls can then be hung in your child's room in their proper sequence.

3. You and your child may wish to work together to create a song to help your child remember the order of the planets from the sun outward. Here's an example which can be sung to the tune of "Row, Row, Row Your Boat":

Mercury, Venus, Earth, and Mars -
All planets in a row.
Revolve around the sun all year,
And that's the way they go.

Fun Fact of the Week

More than 1,000 planets the size of the earth could fit inside the planet Jupiter.

Suggested Children's Books

Lampton, Christopher. *Stars and Planets.* New York: Doubleday, 1988.
Lauber, Patricia. *Journey to the Planets.* New York: Crown, 1987.
Simon, Seymour. *Mars.* New York: Morrow, 1987.
Simon, Seymour. *Saturn.* New York: Morrow, 1985.

Sincerely,

Space Sciences

The Solar System

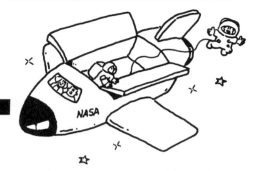

Dear Parents:

Students enjoy learning about the solar system—the planets and their moons, the sun, and the other bodies that move around in space. The solar system includes our sun (which is actually a star), the nine planets—Mercury, Venus, Earth, Mars, Jupiter, Saturn, Uranus, Neptune, and Pluto—and their moons. (Earth is the only planet with just one moon; Saturn, on the other hand, has 17 moons.) The solar system also includes comets (large chunks of ice and dust that orbit the sun), asteroids (rocks that orbit the sun), meteors (small pieces of rocks that enter the earth's atmosphere), and meteorites (any part of a meteor that reaches the surface of the earth). Obviously, there are many objects in our solar system still waiting to be discovered.

The solar system is filled with unlimited opportunities for discovery and exploration. Students are fascinated to learn such facts as:

The surface temperature on Mercury can reach 425° Celsius.
There are two other planets besides Saturn that have rings.
The original name for the planet Uranus was "Herschel."
Thousands of meteorites bombard the earth every year.
It takes Pluto nearly 250 earth years to orbit the sun.

You and your child may wish to select one or two of the following activities for exploration. Be sure to share your findings with other members of the family.

1. Have your child conduct some library research to determine the origins of the names of the planets. For whom were the planets named? Who is responsible for naming the planets? Are there any similarities in the names of the planets? If a new planet were to be discovered today, how would it be named?

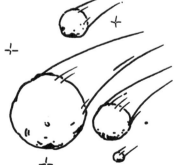

2. Your child might be interested in obtaining some information or printed materials on the travels and discoveries of spacecraft such as the *Voyager* and *Viking* spacecraft. Have him or her write to the NASA Jet Propulsion Laboratory (California Institute of Technology, Pasadena, CA, 91109). Be sure to discuss the information obtained with other family members, too.

3. The first trip from the earth to the moon took approximately four days. Have your child calculate the time it would take for a rocket ship to travel from the earth to each of the nine planets. Use an average rocket speed of 40,000 miles per hour for the calculations. After your child has computed the approximate times, discuss with him or her some of the preparations and provisions that would have to be considered prior to some of the longer manned voyages. What difficulties might be encountered?

Fun Fact of the Week

The tail of a comet can extend for about 90 million miles—or the distance between the sun and earth.

Suggested Children's Books

Clapham, Francis and Ron Taylor (Eds.). *Astronomy Encyclopedia.* New York: Rand McNally, 1984.

Gallant, Roy. *The Macmillan Book of Astronomy.* New York: Macmillan, 1986.

Harris, Alan and Paul Weissman. *The Great Voyager Adventure: A Guided Tour Through the Solar System.* New York: Messner, 1990.

Herbst, Judith. *Sky Above and Worlds Beyond.* New York: Atheneum, 1983.

Sincerely,

Space Sciences

Exploring Space

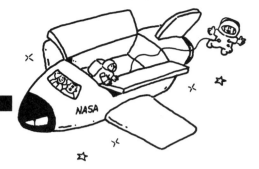

Dear Parents:

Isn't outer space amazing? Distant planets, expanding galaxies, and adventurous voyages by rocket and satellite all add to the excitement. Movies, too, present all manner of possibilities (and impossibilities) about the far reaches of the universe.

When a new space shuttle lifts off or a satellite is launched into space, we can follow the action in newspapers or on television. We can learn much about space from the pictures sent back to earth and the discoveries made on some of these far-reaching voyages. As we learn more about the planets, comets, asteroids, and stars in space we also begin to realize how much we still do not know.

Some of the most amazing scientific discoveries have been found in the farthest reaches of space. Helping children appreciate the potential of future space discoveries is an important part of any science program. You can help your child participate in this aspect of science by selecting one or two of the following activities to share together. You and your child might also create additional activities with the help and assistance of a local high school teacher or college professor. Contact the science department at the school.

1. If possible, purchase or obtain a telescope for your child. A telescope will allow your child to conduct his or her own investigations of the skies and learn first-hand about some of its more amazing features. Before purchasing a telescope, however, you may want to obtain a copy of "Selecting Your First Telescope." This useful publication is available for a small donation to the Astronomical Society of the Pacific (1290 24th Avenue, San Francisco, CA 94122).

2. NASA (the National Aeronautic and Space Administration) has many different brochures, pamphlets, photographs, slides, and tapes available on space exploration. Write to one of the following regional centers for further information: Ames Research Center (Moffet Field, CA 94035); Johnson Space Center (Houston, TX 77058); Kennedy Space Center (Kennedy Space Center, FL 32899); or Langley Research Center (Langley Station, Hampton, VA 23365).

3. Plan to visit a planetarium at a nearby high school or college. Many offer regularly scheduled programs for the public. These programs offer explorers of all ages glimpses into the universe and opportunities to talk with experts about recent space discoveries.

Fun Fact of the Week

To escape the earth's gravitational pull, a spacecraft must travel faster than seven miles per second.

Suggested Children's Books

Allen, Joseph. *Entering Space: An Astronaut's Odyssey.* New York: Workman, 1984.

Branley, Franklyn. *From Sputnik to Space Shuttles: Into the New Space Age.* New York: Crowell, 1986.

Maurer, Richard. *The NOVA Space Explorer's Guide: Where to Go and What to See.* New York: Potter, 1985.

Snowden, Sheila. *The Young Astronomer.* London: Usborne, 1983.

Sincerely,

The Human Body

Body Support and Movement

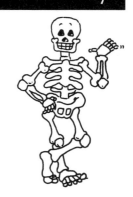

Dear Parents:

Bones are some of the most amazing organs of the human body. Not only do they provide support for the body, but they also protect our other body parts and allow us to move. Connected to those bones are the muscles, which permit movement and add additional strength and structure to the body.

Children use their bones and muscles in physical activities or sports, in performing some chores around the house, or in getting from place to place. Students might also have some misconceptions about these body parts. They are often amazed to discover that girls have the same number of muscles as boys. A professional football player has the same number of muscles as a 10-year old girl (they just happen to be better developed).

Bones are also important for making scientific discoveries about history. Scientists can learn a great deal of information about prehistoric or ancient peoples by examining bones. From bones, scientists can deterine the age, sex, weight, height, mode of death, and general health of a person who lived thousands of years ago.

You can help your child learn more about the structure and function of the bones and muscles in his or her body with one or two of the following activities.

1. If possible, seek permission to visit a local physical rehabilitation center or chiropractor. Encourage your child to talk with the personnel about the structure and function of the human skeleton and muscle system. What are some of the ways in which we can protect our bones and muscles? What are some possible exercises and dietary habits? What happens to people who suffer diseases or injuries to their bones or muscles? On your return home, have your child collect the information together into a brochure to be shared with other family members or taken to class.

2. CAUTION: An adult should supervise this activity. Obtain a whole chicken and boil it for one or two hours. Carefully remove all the meat. With your child, observe the structure of the skeleton. Have your child draw an illustration of the chicken's skeleton. After the skeleton has cooled, work with your child to remove bone sections from the skeleton. Have your child note the joints and how they are held together with **ligaments** (muscles that connect bone to bone).

3. The human skeleton continues to grow until sometime between the ages of 16 and 22. Have your child record the heights of all family members. This can be done once each month. Have your child make predictions about each person's height for the forthcoming month. Which persons in the family are continuing to grow? Who has stopped growing? Your child might want to create a special chart or graph of family members' growth. Friends and other relatives can also be added to the chart.

Fun Fact of the Week

Human skeletal muscles can contract and relax in less than 0.1 seconds. Human heart muscles require one to five seconds to relax and contract.

Suggested Children's Books

Baldwin, Dorothy and Claire Lister. *The Structure of You and Your Body.* New York: Bookwright, 1984.

Elting, Mary. *The Answer Book About You.* New York: Putnam, 1984.

Kapit, Wynn and Lawrence Elson. *The Anatomy Coloring Book.* New York: Harper & Row, 1977.

Showers, Paul. *You Can't Make a Move Without Your Muscles.* New York: Crowell, 1982.

Sincerely,

The Human Body

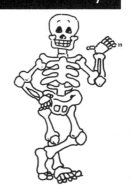

Growing and Changing

Dear Parents:

One of the things we are very aware of is the growth and development of our children. It is not unusual for many parents to keep accurate records of how tall their youngsters are at various ages. Graphs and charts on door jambs and family photo albums are features in many homes. We are concerned when our children do not "measure up" to the other children in the neighborhood or when their height exceeds most of the other children in their grade.

Growth is a very complex process that begins at birth and continues until a child is between 16 and 22 years of age. Girls usually reach their full adult height by the age of 16. Boys reach their full height by about 20. Growth is determined by diet, hormones, and heredity. Having a balanced and nutritious diet positively affects the amount of growth a child will experience over his or her developmental years. Lack of some nutrients can affect a child's eventual height.

Hormones secreted by the **pituitary gland**—a small gland at the base of the brain, are also responsible for growth. The hormones are released in small amounts when a child is young. When a child reaches the ages of 9 to 12, the pituitary gland begins to make larger amounts of the growth hormones. This is the time when youngsters make the largest growth spurts. Heredity also plays a role in a child's growth. If his or her parents are tall, it is likely that the child will be tall as well. If both parents are short, the child will also have a tendency to be short.

As your child grows and develops, you can help him or her appreciate some of the processes taking place by selecting one or two of the following activities. It will be important for your child to understand that all children grow at different rates and in different degrees; again dependent on hormones, diet, and heredity. Your child's growth is different from brothers, sisters, and other children in the neighborhood.

1. When a baby is born, its head is about one-fourth the length of its entire body. As a baby grows, the head becomes smaller in comparison to the rest of the body. Have your child select several photographs of various family members at different ages. Ask your child to measure the length of each family member's head as well as the length of each person's whole body. What kinds of proportions does your child note? What is the proportion of the head to the body in infancy, childhood, teenage years, young adult, and the adult stages? Are there some trends in your family?

2. Look through several old magazines with your child for photographs of young animals. Have your child note the proportion of body parts in several young animals in comparison with the adults. For example, a young horse has legs that seem to be too long, but an adult horse has legs that are just right in proportion to other parts of the body. What other animals have body parts that vary in size from youth to adulthood? What about your dog's paws and ears?

3. Visit your family doctor or a local health clinic to obtain information on the suggested heights and weights of children at different ages. Please keep in mind that these measurements are only averages, and might be different from the height and weight of your child. Talk with a doctor about the diet, sleep, and exercise your child needs to maintain proper growth and development.

Fun Fact of the Week
It takes a human being about 4-1/2 months to replace a fingernail from the quick to the tip.

Suggested Children's Books
Bruun, Ruth and Bertel Bruun. *The Human Body.* New York: Random House, 1982.
Caselli, Giovanni. *The Human Body.* New York: Grosset, 1987.
Cole, Joanna. *How You Were Born.* New York: Morrow, 1984.
Daly, Kathleen. *Body Words: A Dictionary of the Human Body, How it Works, and Some of the Things That Affect Its Health.* New York: Doubleday, 1980.

Sincerely,

The Human Body

Digestion and Circulation

Dear Parents:

The human body is an incredible machine! Certainly one of its most fascinating features is the fact that it carries on several different functions simultaneously without our ever having to think about them. **Digestion** (the changing of food into a form the body cells can use) and **circulation** (the transportation of nutrients throughout the body) are certainly two of the most amazing processes.

Digestion begins when food enters the mouth. There it is chewed and moistened so that it can pass down through the **esophagus** (the tube from the mouth to the stomach) and into the stomach. In the stomach, the food is mixed with juices and acids and passed into the small intestine. From the small intestine, the nutrients in food can pass into the body. The undigested parts of food move into the large intestine, which stores this waste until it leaves the body.

The circulatory system is the body's means of distributing needed materials to body cells and removing wastes from those cells. Driven by the heart, blood filled with nutrients reaches almost all parts of the body. Oxygen, water, nutrients, and wastes pass easily between body cells and the blood in the smallest vessels, known as **capillaries.**

You can help your child understand the functions of these two body systems by using one or two of the following activities.

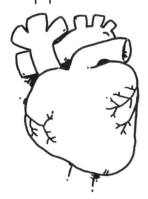

1. If possible, visit a nearby blood bank and ask to talk with one of the technicians or nurses. Find out how blood is collected, measured, stored, and preserved. What precautions do the workers have to follow? How much blood is collected in a day, a week, or a month? How is that blood used? How is it transported? In light of the AIDS epidemic, how is the blood tested and protected?

2. CAUTION: An adult should use the knife in this activity. Visit a local butcher shop and obtain a calf's heart. With your child watching, cut the heart open and observe some of its parts. Compare your findings with an illustration of a human heart in a science book. (A calf's heart is very similar to a human heart.) Have your child draw a picture of the calf's heart and share his or her findings with other family members.

3. In each of three small plastic cups, pour 3 tablespoons of milk. In the first cup put 2 tablespoons of water. Cover the cup with a sheet of plastic wrap, using a rubber band to hold the wrap in place. In the second cup, put 2 tablespoons of a weak acid such as vinegar or lemon juice and cover as above. In the third cup put 2 tablespoons of an enzyme such as meat tenderizer and cover as above. After one to two hours observe the changes that occur in each cup. The changes that occurred in cups 2 and 3 are similar to the digestive process in the stomach.

Fun Fact of the Week

In one year, your heart will beat about 36 million times.

Suggested Children's Books

Ontario Science Centre. *Foodworks: Over 100 Science Activities and Fascinating Facts That Explore the Magic of Food.* Reading, MA: Addison-Wesley, 1987.

Peavey, Linda and Ursula Smith. *Food, Nutrition and You.* New York: Scribner's, 1982.

Silverstein, Alvin and Virginia Silverstein. *Heartbeats: Your Body, Your Heart.* New York: Harper, 1983.

Wilson, Ron. *How the Body Works.* New York: Larousse, 1978.

Sincerely,

The Human Body

The Brain and Sense Organs

Dear Parents:

We don't spend much time thinking about our brains, yet our brains control so much of what we do everyday. Thanks to our brains, we are able to breathe, eat, work, and sleep without really thinking about it. As the control center for our body, our brain takes over the proper functioning of many of our unconscious activities. Obviously, the brain is needed to learn and perform many conscious activities as well.

The brain is truly one of the most unique organs in the human body. Although it weighs about three pounds, it is responsible for controlling what we see, hear, taste, smell, and touch. The brain acts as a "collection agency." It receives nerve impulses from all parts of the body, sorts through them, and sends vast quantities of information back to all parts of the body. What is even more amazing is that although the brain controls all the nerves of the body, it has no nerves at all. In fact, a surgeon can cut deep into a human brain and the patient will not feel any pain!

Students need to understand some of the intricacies of the human brain—what it looks like, how it functions, and what it does. Although we still have much to learn about the human brain, students can begin to appreciate its importance with one or two of these activities.

1. CAUTION: An adult should use the knife in this activity. If possible, visit a local butcher shop and get a calf's brain. As your child watches, cut into the brain and observe some of its major features. (The shape of a calf's brain is similar to that of a human's.) Ask your child to draw an illustration of the brain; when completed compare that picture to an illustration of the human brain found in an encyclopedia or one of the books below.

2. Provide members of the family with five six-digit numbers (for example: 610448, 729505, 265410, 478246, 193072). Ask each member of the family to memorize those numbers as rapidly as possible. Have a brief "quiz" to see who memorizes the numbers best. Have a family discussion on some of the possible reasons why some family members were able to memorize the numbers better than others.

3. Talk to your family doctor or a health care worker about some of the ways in which a person should protect his or her eyes and ears. You might be able to obtain a brochure or other informational guide on the care of these sensory organs. If you can't find one, write to the American Speech-Language-Hearing Association (10801 Rockville Pike, Rockville, MD 20852) and ask for some descriptive literature.

Fun Fact of the Week
A nerve impulse travels to your brain at a speed of 205 MPH.

Suggested Children's Books
Bruun, Ruth D. and Bertel Bruun. *The Brain: What it is, What it Does.* New York: Greenwillow, 1989.

Elting, Mary and Wyler, Rose. *The Answer Book About You.* New York: Grosset & Dunlap, 1980.

Gaskin, John. *The Senses.* New York: Watts, 1985.

Stafford, Patricia. *Your Two Brains.* New York: Atheneum, 1986.

Sincerely,

The Human Body

Respiration and Excretion

Dear Parents:

As we walk around at home or do some shopping at our local shopping malls, we may not be aware of our respiratory and excretory systems. But if we run upstairs or engage in some sort of strenuous physical activity then we become aware of our body's need to take in more oxygen and get rid of some perspiration. Panting and sweating are two signals that our respiration and excretory systems are working.

Respiration is the process in which the human body takes in air and then uses the oxygen from that air. Our body cells need this oxygen in order to function. Air is taken in (or **inhaled**) through the mouth and nose. Then it passes into the **trachea**—the hollow tube from the mouth to the lungs. Inside the lungs, the trachea divides into smaller and smaller tubes that eventually lead to very small air sacs inside the lungs. Oxygen passes through these air sacs into the body. At the same time, carbon dioxide—wastes from the body cells—passes from the blood into the sacs. The blood then takes the oxygen to all parts of the body.

Just as it is important for the body's cells to have oxygen in order to work properly, it's important that they get rid of the wastes—such as water, salts, sugars, nitrogen, and carbon dioxide that the body cannot use. This is done through the **excretory** system. Some of the wastes are carried by the blood to sweat glands in the skin and are released in the form of sweat. Other wastes are removed directly from the blood by the **kidneys**, which are very similar to an air filter in a car. The kidneys remove nitrogen wastes, salts, and extra water from the blood and send them to the **urinary bladder** in the form of **urine**.

You can help your child understand and appreciate these two body systems by selecting one or two of the following activities.

1. Breathing is an action we take for granted. Children should be aware of some of the things they breathe into their bodies. Have your child take several 3" x 5" index cards and tape or tie them to various places in and outside of the house (for example, on the side of the refrigerator, on the ceiling of the living room, outside just above the front door, on a branch of a nearby tree). On each card smear a thin layer of petroleum jelly. After several days have your child note the amount of pollutants on each card. Which card had the most? Why? Check the cards over a span of two or three weeks.

2. Most humans need approximately eight glasses of water a day in order to properly maintain their excretory systems. Although some water is obtained through vegetables, soups, and milk products, most individuals need to drink regular amounts of tap water, too. Have your child construct a chart listing the names of all family members. Ask each family member to report to your child the exact amount of water or number of glasses of water he/she drinks each day. Have your child maintain those records for two or three weeks. Discuss with family members the importance of a constant water supply for the body to function properly.

3. You and your child might want to write or call the National Institute of Health (Building 31, Room 2B19, Bethesda, MD 20892; 301-496-8855) to request information or literature on some of the diseases of the lungs and kidneys as well as research being conducted to prevent or reduce the dangers of those diseases. Share the information with all family members.

Fun Fact of the Week
The average person takes approximately 23,040 breaths every day.

Suggested Children's Books
Baldwin, Dorothy and Claire Lister. *The Structure of You and Your Body.* New York: Bookwright, 1984.

Elting, Mary. *The Answer Book About You.* New York: Putnam, 1984.

Kramer, Stephen. *Getting Oxygen: What to Do if You're Cell Twenty-two.* New York: Crowell, 1986.

Miller, Jonathan. *The Human Body.* New York: Viking, 1983.

Sincerely,

The Human Body

Staying Healthy

Dear Parents:

Here's an interesting question to ask your child: Why is it important to stay healthy? While the answer to that question may be obvious to us as adults, you may discover that your youngster has some degree of difficulty in responding to that particular question. Too often we ignore our health until we experience a disease, illness, or other medical emergency. What is most important for children is the fact that their individual health is something that should concern them on a regular basis. Also important is the attitude that each person is directly responsible for his or her own health.

The attitudes and practices children develop early in their lives will have a profound effect on their health as adults. It is therefore important that parents provide information and support in helping their youngsters develop healthy lifestyles. Given the increasing cost of medical care in this country, it is vital that children become aware of the practices that may prevent or eliminate later health problems. Regular visits to a family doctor and dentist, a healthy and nutritious diet, and regular exercise are all critical in helping youngsters grow and develop. They also help form the basis for their continued health throughout their lives.

We shouldn't take our children's health lightly—it will determine the kind of life they will enjoy for many years to come. Please plan to share one or two of the following activities with your child. Also, plan time to share some thoughts and ideas about a healthy lifestyle with other members of the family.

1. Many family doctors, hospitals, and health clinics have a variety of brochures and other printed information on health and nutrition that you can pick up for your family. Take some time to share and talk about the information in these printed materials.

2. The three major health needs for growing children are a nutritious diet, regular exercise, and sufficient sleep. Have your child make a personal notebook divided into three sections. In the first section have your child record the foods he or she eats (even snacks) during the course of a week. Your child might want to divide a sheet of paper into the four major food groups ("vegetable-fruit group," "bread-cereal group," "meat-poultry-fish-bean group," and "milk-cheese group") and record the types and amounts of each eaten during the week. In the next section have your child record the type and duration of various physical activities he or she participates in during the week. In the third section have your child record the number of hours of sleep he or she get during each night of the selected week. Your child might want to share his or her notebook with a family doctor for comments and suggestions.

3. Take time to talk with your child about the dangers of alcohol, tobacco, and drugs. The attitudes children have about these substances are formed very early in childhood so it is vitally important that you discuss some of the problems associated with these substances. Important information and materials can be obtained from your local hospital, health clinic, or family doctor. Or contact the National Clearinghouse for Alcohol and Drug Information (P.O. Box 2345, Rockville, MD 20852, 800-729-6686) for a free catalog of their materials. This is one activity that you'll want to share with your child throughout this year and the years to come.

Fun Fact of the Week
43 percent of the human body is muscle.

Suggested Children's Books
Berger, Melvin. *Why I Cough, Sneeze, Shiver, Hiccup and Yawn.* New York: Crowell, 1983.
Burnstein, John. *Slim Goodbody: What Can Go Wrong and How to Be Strong.* New York: McGraw-Hill, 1978.
Nourse, Alan. *Your Immune System.* New York: Watts, 1989.
Settel, Joanne and Nancy Baggett. *Why Does My Nose Run? And Other Questions Kids Ask About Their Bodies.* New York: Atheneum, 1985.

Sincerely,

Activity Calendars

These reproducible calendars provide families with a wealth of exciting and interesting science activities to do throughout the year. There are books to read, places to go, experiments to conduct, people to talk to, and questions to answer. All in all, there is something for everyone!

Please note that there are some "blanks" for each month. This allows you an opportunity to write in some of *your own* activities (prior to reproducing the calendars) for parents and children to share. Select certain activities from the science text, a teacher resource book, a professional publication, or from your own imagination. Or you may want to write in "Free Time" in some of these blank spaces to allow families an opportunity to select or create their own science-related activities.

September

Sunday	Monday	Tuesday	Wednesday	Thursday	Friday	Saturday
Read a science book together.	How many days does it take the earth to make one revolution around the sun? Do some library research and share the answer with family members.	Find out how long the San Andreas Fault in California is. Draw an illustration of the fault.	Write a letter to the editor of your local newspaper about the destruction of the Amazon rain forest.	Which is most dangerous—hurricanes, tornadoes, earthquakes, or volcanoes? Look in the library and discuss your findings with family members.		**Adult Supervision Needed.** Stretch a balloon to loosen. Stretch it over the mouth of a bottle. Place the bottle in a pan of water on the stove. Heat the water. What happens?
Read a science book together.	What is the difference between "populations" and "communities" of animals?	A person's small intestine is about 1-1/2 times longer than his or her height. Calculate the lengths of each family member's small intestine.	Scatter bird seed in an area around the outside of your house. How many different kinds of birds come to eat the seed?	Make a list of some of the common elements used in your house. Are there more elements in the kitchen, living room, garage, or bathroom?	What are some examples of weathering? What are some examples of weathering in your local area?	Brainstorm with your family about some of the sounds you normally hear in your house. Are they the same or different from those sounds heard in a friend's house?
Read a science book together.		What are some of the most active volcanoes in the world? Where are they located?	Write to Edmund Scientific (101 E. Gloucester Pike, Barrington, NJ 08007) and ask for a copy of their latest catalog.	Work with family members to make a list of the different types of mechanical, chemical, and electrical energy used in your home.	What does your liver do? Visit the library and find out. Share the information with family members.	Visit a pet store and obtain some mealworms. Place them on a sheet of graph paper and note the kinds of movements they make. Trace their movements with a pencil.
Read a science book together.	Read David Adler's book *Amazing Magnets*. Can you create your own magnet at home?		Write a newspaper article about how to conserve electricity.	Who or what is the biggest polluter in your area? Make a list of three things that can be done to reduce that kind of pollution.	If you count the number of seconds between seeing lightning and hearing the thunder, and divide by three, you will know how far away the lightning strike is (in kilometers).	Fill several different sized bottles with varying levels of water. Blow across the top of each to create a sound. Which bottle produced the sound with the highest pitch? Why?
Read a science book together.	Find out what a *tsunami* is. Where do they usually occur?	Get some Styrofoam balls and strong wire from a local hobby store. Work with family members to create a model of an atom.		Collect several examples of leaves and arrange them in various categories (e.g., pointed, smooth, etc.).	For a catalog devoted solely to dinosaur-related items write to "Dinosaur Catalog" (P.O. Box 546, Tallman, NY 10982).	Obtain some sandstone from a local building supply dealer. Break the stone into pieces with a hammer. How is that process similar to events on the surface of the earth?

October

Sunday	Monday	Tuesday	Wednesday	Thursday	Friday	Saturday
Read a science book together.	Sprinkle a piece of bread with water and place it in a sealable sandwich bag. Keep it on a windowsill. What happens to the bread after several days?	Draw a map of your house or yard. Using a compass, point "north," "west," "south," and "east."	Make a list of all the different kinds of desserts that can be made with apples.		Is a 100-watt light bulb brighter than a 25-watt light bulb? How can you find out?	Get several different pieces of glass (be careful) and hold them in such a way as to create a rainbow. What colors are in a rainbow?
Read a science book together.	Obtain several different kinds of plants. Cut the stems in half lengthwise and crosswise. What similarities do you note?	Stretch three rubber bands of different sizes. Which one has the most potential energy? Which one has the most kinetic energy? How can you find out?		List four ways wild animals get their food. Which way is most common?	Find out where meteorites come from and what they are made of.	Collect several different kinds of rocks. Arrange them into an attractive display.
Read a science book together.	Take a house plant and tape small pieces of paper over parts of several leaves. After several days, remove the paper pieces. What has happened? Why?	What kind of surface will absorb the most heat—blacktop, grass, or a wood deck? How can you find out?	Do some library research. List two ways in which energy changes form. What examples can you find in your home?	Locate two different kinds of flying insects in your neighborhood. What similarities do you note between the two?		Fill a cake pan with soil. Tilt the pan at different angles. Let water from a bottle flow from the top of the pan to the bottom. At which angle does erosion cause the most damage?
Read a science book together.	Draw an illustration of a plant cell. What are some of the major parts?		Rub two different grades of sandpaper together. Which one wears the most? Why? What does this mean in nature?	Write a commercial for your favorite vegetable.	If an earthquake measures 6 on the Richter scale, how many times stronger is it than an earthquake that measures 4 on the scale?	Get an old telephone from a yard or garage sale and take it apart. Explain to family members the function of some of the parts.
Read a science book together.	Explain the processes of evaporation and condensation to a younger family member. How could you demonstrate these processes?	Look at some chicken bones. Look at some beef bones. What similarities do you note? What differences do you see?		Why is it important for some animals to live together in large groups? Do some library research and list three main reasons.	Read *Introduction to Physics* by Amanda Kent and Alan Ward. List some ways in which you use sound, light, and magnetism in your daily life.	Find out what **epiphytes** are. Where in the world would you expect to find them?

November

Sunday	Monday	Tuesday	Wednesday	Thursday	Friday	Saturday
Read a science book together.	How long does a lightning bolt last? Where could you go to locate that information?	Who has more muscles—boys or girls? Do some research in the library and share your findings with family members.	Make a list of good conductors of electricity. Make another list of poor conductors of electricity. Which list is longer? Why?	What are **monerans**? Where would you expect to find them? Would you find them in your home?		Research all the different uses that Native Americans had for the buffalo.
Read a science book together.	Find pulse rates of family members and record them. Test members after they do work or physical activity. Whose rate differs the most between "rest stage" and "active stage"?	Read *What's Under That Rock?* by Stephen Hoffman. What kinds of animals would you expect to find under the rocks around your home?	Talk with family members about some of the dangers of relying on fossil fuels as a source of energy.		Develop a plan for reducing or eliminating acid rain. Discuss your plan with family members.	Visit a local blood bank and talk to one of the technicians about methods used to collect, test, and store blood.
Read a science book together.	Fill a cereal bowl with dry oatmeal and a few of paper clips. Have someone blindfold you. How many paper clips can you remove from the bowl in one minute?	Look in the library for some examples of food chains. Create an illustration of a food chain that may take place in your area.	Write a letter to the editor of a local newspaper about an important environmental issue.	What are the differences between herbivores, carnivores, and omnivores? Draw a picture of an example of each.	Write a newspaper article about the life and death of a tree.	Will a hard-boiled egg float better in salt water or fresh water? Can you design an experiment to find out?
Read a science book together.	Describe the habitats of at least four different animals that live in or around your neighborhood.	Draw a picture of the human eye and the path light travels when it reaches the eye.		Do bodies of water freeze from the top down or from the bottom up? How can you find out?	Talk with family members and friends about their favorite animals. Which one is the most popular? Why?	What kind of energy is there in the batteries in your flashlight? How is that energy turned into light energy?
Read a science book together.	Write to JETS (Junior Engineering Technical Society, 345 East 47th St., New York, NY 10017) and ask for information about their science clubs and newsletter.		Create a poster that shows the different phases of the moon.	Name two ways in which plants adapt to cold climates. Name two ways in which animals adapt to cold climates.	What is the most poisonous animal in the world? What is the most poisonous plant?	Which city—Los Angeles, Seattle, Phoenix, or Denver—gets the most precipitation? How can you explain the differences?

December

Sunday	Monday	Tuesday	Wednesday	Thursday	Friday	Saturday
Read a science book together.	Write to the U.S. Department of Energy (Technical Information Center, P.O. Box 62, Oak Ridge, TN 37830) and ask for a free copy of "Science Activities in Energy."	What are some of the ways in which geothermal energy is used in the United States? Share your information with family members.	Over the course of several days, look in your newspaper and record the barometric pressure. Why does the barometric pressure differ from day to day?	Read *Is There Life in Outer Space?* by Franklyn Branley. Do you think there is life on other planets? Why?		Blow up a round balloon. Dip newspaper strips in a bowl of liquid starch and cover the balloon. When the strips are dry, paint the balloon like a globe.
Read a science book together.	Write a poem about electricity. Share it with family members or ask each member to add a new line.	What is the star closest to the earth? How far away is it? How long would it take to reach that star if you could travel at 25,000 MPH?	List some of the voluntary muscles in your body. List some of the involuntary muscles. Which ones are used most?	What are some of the characteristics of winter in your part of the country? How do those differ from a place 1000 miles away (north, south, east, or west)?	Where are some of the major coal-producing regions in this country? Where are some of the major oil and natural gas regions? Do you note any similarities?	Make up a week-long menu of nutritious foods. Help prepare some of the foods for the family.
Read a science book together.		Put out a bird feeder and count the number of different types of birds that visit it during the course of a week.	Find out the difference between whole milk, 2 percent milk, and skim milk. Share the information with family members.	Which is the most serious issue—water pollution, air pollution, or land pollution? Discuss your opinions with family members.	Read about Thomas Edison. How many different inventions did he patent in his lifetime? Which one do you think was the most important?	Visit the local public library and check out a video about tropical rain forests.
Read a science book together.	Make a list of the different kinds of work you do each day. Which one requires the most energy? Why?	What is a **joule**? Can you describe to a family member what it means?		Do some library research and create an exercise program for each member of the family.	List reasons why it is important to recycle and reuse. Share the list with family members.	Wet some steel wool and place it in a sealed bottle. What happens to it after a week or so?
Read a science book together.	Create a poster that describes the process of metamorphosis.	What is a **vernal equinox**? When does it occur?	Obtain a sweet potato and cut it in half. Push toothpicks around the side and suspend it in a glass of water, cut side down. What happens to the potato after a few days?	What is the largest land animal? Largest sea animal? Largest air animal? How do they compare in weights?	Read *Voyage of the Ruslan: The First Manned Exploration of Mars* by Joshua Stoff. Discuss why you would like to live on Mars.	

January

Sunday	Monday	Tuesday	Wednesday	Thursday	Friday	Saturday
Read a science book together.	Find out how many different varieties of cherries (or peaches, apples, or plums) there are.	Read *Simple Electrical Devices* by Martin Gutnik. Create one of the devices explained in the book.	Get three thermometers and place one near the ceiling of a room, one near the middle, and one near the floor. How do you account for the differences in temperature readings?	Write a letter to the editor of the local newspaper on why families should recycle.	There are only two mammals in the world that lay eggs. Find out what they are and share your information with family members.	Do some library research about extinct animals. How many animals become extinct each year?
Read a science book together.	Explain to family members how an electromagnet works. How are electromagnets used in manufacturing and industry?	Draw a large chart that shows how blood flows through the human heart. Explain your drawing to family members.	Read *Blood and Guts: A Working Guide to Your Own Insides* by Linda Allison. What was the most amazing thing you learned as you read this book?	Obtain a National Geographic video from your local video store and watch it together.	Read about how spiders construct their webs. Can you do the same thing with string? Try it.	
Read a science book together.	What are some of the ways in which distances in space are measured? Draw a chart or graph and explain it to family members.	Make a chart with the words "Plastic," "Paper," "Metal," and "Food" listed across the top. Record the quantities of each your family throws away.	Stars may be blue, blue-white, green, yellow, red, or orange in color. Which colors denote hot stars and which colors denote cool stars?	Explain to a family member how the moon gets its light.	Use a hinge, some pieces of wood, and several rubber bands to construct a model of a human muscle at work.	The Bay of Fundy is noted for a particular phenomenon. What is it?
Read a science book together.		Make a list of ten ways to keep warm in winter. Which one do you like best?	Read *Scavengers and Decomposers: The Cleanup Crew* by Pat Hughey. Share some of the information with family members.	Interview an older relative and discuss some of the electrical appliances he/she did not have when growing up.	Why is it colder at the top of a mountain than it is at the bottom of a mountain?	**Adult Supervision Needed.** Put on an oven mitt and hold a cookie sheet over a kettle of boiling water. (Be careful!) Why do water drops form on the bottom of the cookie sheet?
Read a science book together.	What is **convection**? Draw an illustration to show how convection works in your home.	Go through your kitchen and make a list of all the simple machines you can find.		Read *101 Questions and Answers About the Universe* by Roy Gallant. Can you think of a question about the universe that was not answered in the book?	Write a commercial for your favorite kind of weather.	What is the weather like in Australia? Would you rather be in Australia or the U.S. right now?

February

 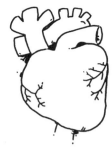

Sunday	Monday	Tuesday	Wednesday	Thursday	Friday	Saturday
Read a science book together.	Explain the differences between reptiles and amphibians to members of your family.	Why can't mercury thermometers be used at temperatures below $-39°C$ ($-38.2°F$)?	How can some plants reproduce without seeds? Do some library research and share your findings with family.	Make a list of some of the endangered species (plants or animals) that live in your area of the country. What can you do to help save them?	Read *A Day in the Life of a Marine Biologist* by William Jasperson. Explain why you would like to become a marine biologist.	Which falls fastest—a golf ball, a table tennis ball, a tennis ball, or a baseball? How can you set up an experiment to find out?
Read a science book together.	Tell friends that you can hold 100 pounds in your hand for as long as necessary. Put your hand out, palm up, and you've done it (the weight of air on your hand is about 100 pounds)!	Write to Delta Education, Inc. (P.O. Box 950, Hudson, NH 03051) and ask for a copy of their latest catalog.	If you could have a dinosaur as a pet, which one would you choose? Write a story or poem about your pet.	Which do you support—more emphasis on solar energy or atomic energy? Write an article supporting your view.		What is the difference between camels and dromedaries? Find the answer at your local public library.
Read a science book together.	Place an ice cube in a glass of water. Lay the end of a piece of string on the cube and sprinkle salt over it. Wait a few seconds and lift the other end of the string. What happens?	Write to the National Audubon Society (950 Third Ave., New York, NY 10022) and ask for information about their ecology camps.	What is a **Red Dwarf**? Share your findings with family members.	Read *The New Illustrated Dinosaur Dictionary* by Helen Sattler. Which dinosaur is your favorite? Why?	Using balls and string, create a model of an atom. Be sure to include electron(s), proton(s), and neutron(s).	**Adult Supervision Needed.** Create an instrument using pieces of wood with nails pounded into it and different lengths of rubber bands stretched between the nails.
Read a science book together.		Make a list of all the different kinds of energy used in your home. What kinds are the easiest to conserve?	Find out about heart transplants. When was the first one done? How many are done in this country every year? How dangerous are they?	How many devices in your house produce heat? Make a list and share it with family members.	Survey your community for signs of weathering. Look at buildings, statues, and open areas of land. Where is most weathering taking place?	Place wet paper towels around the inside of a glass jar. Place bean seeds between the towels and the glass. Place the jar in the sun. What happens over several days?
What is a leap year? Why do we have leap years? When will the next leap year occur?		Which is fastest, the speed of light or the speed of sound?	Read *Biological Clocks* by Sarah Riedman. What kind of "clocks" do you follow during the course of the day?	Make a list of some of the changes that happen to your body as you get older. Discuss the list with an adult.	Take a small plant from its container. Tie two wet sponges around the root system and hang the plant upside down in a window. What happens after several days? (Keep the sponges moist.)	Get several pieces of plywood or thin plastic and demonstrate to your family how the plates of the earth move and shift.

March

Sunday	Monday	Tuesday	Wednesday	Thursday	Friday	Saturday
Read a science book together.	The hottest temperature in the U.S. was recorded at Death Valley, California. What was it and when did it occur?	What is the difference between mass and density? How can each one be measured?	Write a dialogue between yourself and a space shuttle astronaut.		What kinds of energy are involved in a roller coaster ride?	Make up a comic strip about the coldest day or week of winter.
Read a science book together.	Read Vicki Cobb's books *Science Experiments You Can Eat* or *More Science Experiments You Can Eat*. Create one of the "experiments" and share it with your family.	Find out where green plants get their energy. Share the information with family members.	Use some modeling clay and create a topographical model of your area of the country.	Would you rather be a fish, reptile, amphibian, or bird? Explain the reasons for your choice.	Make a poster of the nine planets and their orbits around the sun.	List some of the different ways humans use rocks and soil. How do you use rocks and soil in your family?
Read a science book together.		Monarch butterflies migrate between Mexico and the U.S. each year. Trace their migration on a map of the world. Why are monarchs becoming endangered?	Find out how the telephone was invented. What would life be like without the telephone?	Explain to a family member how radar or sonar works.	Read *Our Wild Wetlands* by Sheila Cowing. How do you think we should preserve wetland areas?	Visit a local factory or manufacturing plant. What kinds of tools or simple machines are used?
Read a science book together.	Arthropods are the largest group of animals in the world. Make a list of arthropods found in your local area.		Explain how the human digestive system works. Draw a picture or illustration of the different processes involved.	What is the largest glacier in the world? What was the largest iceberg?	Write to Ranger Rick Wildlife Camp (National Wildlife Federation, 1412 Sixteenth St., Washington, DC 20036) and ask for information about their special summer camps.	Obtain an ink pad from a local office supply or stationery store. Have each family member make fingerprints of each finger. What differences do you note in the fingerprints?
Read a science book together.	Turn on your cold water faucet so that a thin stream of water comes out. Comb your hair with a plastic comb. Hold the teeth of the comb close to the water stream.	Cut out weather predictions from several different days of your local newspaper. Talk with family members about why it might be difficult to accurately predict the weather.		**Adult Supervision Needed.** Dip a toothpick in lemon juice and write your name on a sheet of paper. Carefully hold the paper over a candle flame and watch what happens.	If you could ask three questions of any scientist in the world, what questions would you want answered?	Explain to family members how fossils form. Where would you be able to locate a large number of fossils?

April

Sunday	Monday	Tuesday	Wednesday	Thursday	Friday	Saturday
Read a science book together.	Read *Test-Tube Mysteries* by Gail Haines. Share two amazing facts from the book with family members.	Place one thermometer in the sun and another in the shade. After 30 minutes read each one. How do you account for the different readings?	Write to the Alabama Space and Rocket Center (Tranquility Base, Huntsville, AL 35807) and ask for information about the Space and Rocket Center Youth Science Program.	What are some of the signs of spring? How do plants and animals "know" it's spring? Look in the library to find out.		Is there a zoo nearby? Visit it and ask to see some of the new animal babies.
Read a science book together.	Look around your neighborhood or community for examples of each of the three major kinds of rocks.	Make a list of the major ways in which paper can be recycled.	Discuss with family members what you would most like to change about your neighborhood or community.	Many insects have compound eyes. Draw an illustration of a compound eye.	Write a letter to the editor of your local newspaper suggesting some ways in which families can conserve water.	
Read a science book together.	Write a newspaper article about the severest form of weather that ever occurred in your area.	Is the height of the highest mountain more than the depth of the deepest part of the ocean?	Stretch some rubber bands. Put them in the freezer for 24-hours. What changes occurred in the rubber bands?	Obtain several radish seeds and plant them in plastic cups filled with potting soil. Grow them to maturity.	What is the most dangerous bird, insect, and reptile in the world?	Find out why metal conducts heat better than glass. Can you set up an experiment to prove your ideas?
Read a science book together.		Write a story about "My life as a bug."	Read *How to Be an Ocean Scientist in Your Own Home* by Seymour Simon. What would be the most exciting part of an ocean scientist's life?	What are the two lightest elements in the universe? Where are they found?	The tallest living things in the world are the redwood trees of northern California. How high is the tallest one?	**Adult Supervision Needed.** Using an eye dropper, drop food coloring in a glass bowl of hot water. Drop some in a glass bowl of cold water. What differences do you note?
Read a science book together.	Use a baseball and a bat to demonstrate to family members the difference between potential and kinetic energy.	Write to NASCO (901 Janesville Ave., Fort Atkinson, WI 53538) and ask for a copy of their latest catalog.		Make a bark rubbing of a nearby tree. Place a sheet of typing paper on the surface of the tree and rub over it with a #2 pencil or the side of a black crayon.	Explain to an adult how currents move through the ocean. On a map, point out the flow of the Gulf Stream.	Use different colors of modeling clay and construct a model of the earth's layers.

May

Sunday	Monday	Tuesday	Wednesday	Thursday	Friday	Saturday
Read a science book together.	Write to Leading Edge Boomerangs (51 Troy Rd., Delaware, OH 43015). Ask for their latest catalog as well as information on their newsletter, *The Leading Edge*.	Do you live closer to the North Pole, the South Pole, the International Date Line, or the Equator? How can you find out?	Obtain some seeds from a garden shop. Make predictions as to which seeds will germinate fastest. Plant the seeds and compare their growth against your predictions.	Read John Wexo's *Endangered Animals*. What are some things you and your friends could do to protect wild animals?	Try and locate the Big Dipper in the night sky. How many stars make up the Big Dipper?	Build a bird house and place it outside near a window.
Read a science book together.	What are some of the largest vegetables ever grown? Look in the *Guiness Book of World Records* to find out.		Place several different colored sheets of construction paper in the sun. Which colors will fade the most after 5-6 hours in the sun? Dark colors or light colors?	Read *Uranus* by Seymour Simon. What did you like about it?	Make a terrarium out of a large glass cookie jar or an old aquarium (obtainable at yard or garage sales). Fill it with native plants. What animals will live in this environment?	What is the longest bone in the human body? What is the shortest bone? Where are they located?
Read a science book together.	Where would you most likely find examples of radiant energy in your community or neighborhood?	Read *Castle* by David Macauley. What would you enjoy most about living in a castle? What would you find most difficult?	Find out how the telephone was invented. What would life be like without the telephone?	Read a gardening book together. What kinds of plants would you like to include in your "ideal" garden?		Dig for worms in your yard or a nearby area. Research some of the benefits of worms for gardeners and farmers.
Read a science book together.	Place strips of two-sided cellophane tape on an index card. Place the card outside your house and leave for several days. How much air pollution did the tape collect during that period?	Read about the life of Albert Einstein. What did you enjoy most about his life?	Read and talk about Seymour Simon's book *Saturn*.	Write a commercial for your favorite simple machine.	Read to find out about fish that can live and breathe on land.	Carefully remove a flower from your garden. Lay it on a piece of paper and label as many parts as you can. Consult a reference book, if necessary.
Read a science book together.	Write to Nature Company (P.O. Box 7137, Berkeley, CA 94707) and ask for a copy of their latest catalog.	Make a small rock garden outside your house. Research a book in the public library for ideas.		How long does it take sunlight to reach the surface of the earth? How long does it take for sunlight to reach the surface of Mars?	Take photographs or draw pictures of different kinds of animals with more than four legs. Put the photos together in an album.	Find three fascinating facts about ants and share them with family members.

June

Sunday	Monday	Tuesday	Wednesday	Thursday	Friday	Saturday
Read a science book together.	What are marsupials? Where are most of the world's marsupials found?	Read *How to Be an Inventor* by Harvey Weiss. What kinds of things would you like to invent?	Weed someone's garden. How many different kinds of weeds can you find? Research the weeds that are common to your area.	Start a rock collection. How many different types of rocks can you and your family find in your local area?	Read about water safety. Share the information with family members.	Draw a picture of how sound travels through the human ear.
Read a science book together.		What are some ways in which humans can protect their skin from rays of the sun? Talk about some possibilities with family members.	How many uses do we have for water?	Read *Steven Caney's Invention Book* by Steven Caney. What are some things that you would like to invent?	Make a list of all the different ways seeds travel. Which way is most common? Which way is most unusual?	Find a spider, either indoors or outdoors. Observe its habits for 30 minutes. What are some of the things spiders do?
Read a science book together.	Look in the local newspaper for an article about the weather. Read it to the family.		Visit your yard and collect three different types of bugs. Research the bugs in a book from the library.	Read or get a copy of Vicki Cobb's book *Secret Life of Hardware: A Science Experiment Book*. Select several of the experiments to do with your family.	Explain to your family the difference between weather and climate.	Explain to a family member how a camera works. Or visit a camera store and ask someone to explain the process to you.
Read a science book together.	What is a **galaxy**? Explain your answer to a family member.	Pull up several different kinds of plants. Look at their roots. What do the roots have in common? How are they different?	Make a list of some of the ways in which arteries become clogged.	What is the most important thing we have learned from our exploration of space? Share your thoughts with family members.	Design a postage stamp with a picture of your favorite scientist.	
Read a science book together.	If you could live on any planet in the solar system, which one would it be? Write an article about life on that planet.	Research in the library about how pearls are formed in oysters. Share the information with family members.	A flea can jump 200 times its own length. How far can you jump? Can you jump more than one times your own height?	Read Carol Lerner's book *Moonseed and Mistletoe: A Book of Poisonous Wild Plants*. What are some examples of poisonous plants found in your area?		Write to the Discovery Shop (American Museum of Science and Energy, 300 S. Tulane Ave., Oak Ridge, TN 37830) and ask for their latest catalog.

July

Sunday	Monday	Tuesday	Wednesday	Thursday	Friday	Saturday
Read a science book together.	Look at the night sky and try to locate at least three constellations. Draw an illustration of the constellations you find. Can you invent and draw a constellation of your own?	Bury some potato peelings, newspaper strips, and plastic bottles in different spots in your garden or yard. Dig them up in a month. Which ones have decomposed the most?	What is the hardest metal in the world? What is the softest metal?		Make a poster of the most unusual plant in your neighborhood. Do some library research and include some facts on your poster.	Read *Tree Flowers* by Millicent Selsam. What are some examples of tree flowers in your community or neighborhood?
Read a science book together.	Create three different types of paper airplanes. Can you explain why one flies farther than the others?	Draw an illustration of your tongue and the location of the taste buds for bitter, sour, salty, and sweet.	Read *Dinosaurs, Asteroids, and Superstars: Why the Dinosaurs Disappeared* by Franklyn Branley. Why do you think the dinosaurs died out?	Write an article on why you would like to live at the South Pole. What kinds of adaptations would you have to make in your everyday lifestyle?		Set up a "Bug Tour" or "Bird Tour" of your local neighborhood. Show others some of the usual or unusual animals that live in their area.
Read a science book together.	Read *Safe and Simple Electrical Experiments* by Rudolf Graf. Conduct and explain an experiment on static electricity for your family.	What were some of the tools that cave people used? Can you "invent" one of those tools from materials in your home?	What do you consider to be the most important scientific invention of all time? Why?	If you could be any kind of scientist, what would you like to specialize in or study?	Collect five different varieties of flowers. What characteristics do they have in common?	
Read a science book together.	Write a letter to a friend about the tallest undersea mountain.	Read *How it Feels to Fight For Your Life* by Jill Krementz. Share your feelings about this book with other family members.	Draw a diagram of fossil fuel formation.	Explain to family members the difference between a closed circuit and an open circuit.	Do you know what the largest muscle in the human body is? How can you find out?	In summer electrical and telephone lines seem to sag a little. Why is that?
Read a science book together.		Find the latitude and longitude of the place where you were born.	Explain to a family member why you would like to live on a space station circling the earth.	What is **Absolute Zero**?	Make a list of all the simple machines you can find on a bicycle.	Read the book *Volcano Weather: The Story of 1816, The Year Without a Summer* by Henry and Elizabeth Stommel. What would you do if you had a year without a summer?

August

Sunday	Monday	Tuesday	Wednesday	Thursday	Friday	Saturday
Read a science book together.	Where would you find a **tenrac**? Would you want to have one in your home?	What is the coldest temperature ever recorded on the surface of the earth? Where did it occur?	Which is the best conductor of electricity—copper, aluminum, or iron wire?	Read the National Geographic book *How Things are Made*. Share information about two items in the book with family members.		What do you consider to be the strangest looking animal in the world? Why?
Read a science book together.	What was the longest lunar eclipse? What was the longest solar eclipse?	Draw maps of your garden at home, the garden of a friend, and a neighbor's garden. What similarities do you note?	Find out what kinds of foods vegetarians eat. What kinds of foods would you have to eliminate from your diet to become a vegetarian?	What is the most common type of metamorphic rock? What are some examples of metamorphic rock found in your neighborhood?	How many different kinds of colors do roses have? What colors are not found in the rose family?	
Read a science book together.	Read *Fish Facts and Bird Brains: Animal Intelligence* by Helen Sattler. What is the most amazing thing you learned in this book?	Oysters and clams are similar creatures. Yet they have some important differences. What are some of those differences?	Estimate the height of the tallest tree in your neighborhood. How could you find the actual height?	What is the largest desert in the world? Where is it located?	What do the letters and number mean in the formula H_2O?	How long does ocean kelp grow? Measure and cut a length of string the distance of the longest piece of ocean kelp.
Read a science book together.		Should humans establish colonies on the moon? Discuss your thoughts with other family members.	Take a walk through your neighborhood and locate two plants that you have never seen before. Can you get a book from the library to identify them?	What is the largest planet in the solar system? How much larger than the earth is it?	Read *Discovering Life on Earth* by David Attenborough. Share with family members what you enjoyed most about the book.	Explain to a family member how fish breathe.
Read a science book together.		Find out how many calories are in some of your favorite snack foods.	Give each family member identical balloons. Ask each person to breathe into his or her balloon. Which person has the greatest lung capacity? Why?	What are some of the ways in which humans can change the lives of plants and animals? Discuss with family members different ways of preserving plant and animal life.	Make a list of ten different ways to cool off in the heat. Which one do you use most often?	Get the book *Science Brainstretchers* by Anthony D. Fredericks. Work on puzzles with family members. Which were the toughest? The easiest?

Theme Newsletters

These newsletters will provide opportunities for families to talk about the various science-related themes shared in your classroom as well as to participate together in a variety of science activities.

Each newsletter is an "instant publication," to be completed by the families at home. You may wish to provide opportunities for students to share their completed newsletters in class either through bulletin board displays or learning centers.

Each of the nine newsletters is designed to be used at any time during the course of the school year. To allow you to send your own personal note to families, the pages have been designed so that you can place a 3" by 5" index card with your message on it over the "Dear Family" section before you photocopy. Of course, you are also encouraged to add additional "sections" to any newsletter in keeping with your classroom science program or design.

Growing Green: The World of Plants

Activities:

- Take a walk around your house and find five different types of plants.

- Make a list of all the plants eaten in one day.

- Make another list of the plants used in your family's clothing.

- Take photographs of different plants in your neighborhood.

Dear Family:
Your child will be studying the wonderful world of plants—how they grow, how they flower, and how they survive. Your entire family might want to collect pictures of various plants and put them together in a special "dictionary."

- Write a poem about your favorite plant.

Books for Family Reading:

In The Forest: A Portfolio of Paintings
by Jim Arnosky. A wonderful collection of paintings of the plant and animal life of the forest.

Journey Through a Tropical Jungle
by Adrian Forsyth. This book offers an inspiring argument for the preservation of the world's tropical regions.

Flowers for Everyone
by Dorothy Hinshaw. A delightful exploration of the life cycle of various flowers complete with colorful illustrations.

Growing Green: The World of Plants

Some of my favorite plants

- Plants I eat, but never wear
- The most unusual plant I have ever seen or heard about
- Ways in which plants are harmful to humans
- Plants that can be found in a garden
- Plants that are not green
- Some of my favorite flowers

Fur and Fins: The Lives of Animals

Dear Family:
Your child will be studying all kinds of animals—from the largest dinosaurs to the smallest insects. Your family might want to visit a zoo, aquarium, or animal preserve and talk about the wide variety of animals in the world today.

Activities:

- Research a favorite animal.
- Check the newspaper for stories about animals.
- Prepare a guide on the care of a favorite pet.
- Make a list of the best animals to have as pets.
- Some of my favorite animals.
- Take photographs of animals in your neighborhood or town.

Books for Family Reading

101 Questions and Answers About Pets and People by Ann Squire. A book full of questions and answers about some of the common and not-so-common pets people keep in their homes.

Amazing Poisonous Animals by Alexandra Parsons. One of a series of books about some remarkable creatures in the animal kingdom.

Insect Metamorphosis: From Egg to Adult by Ron and Nancy Goor. Amazing photographs highlight this book about the lives of familiar insects.

Fur and Fins: The Lives of Animals

Animals that live in the desert

Animals that live in our house

The most unusual animal

The last animal I would ever want to meet:

If I could be any animal for a day . . .

Features animals have, but humans don't

Living Things Need Each Other: Ecology & Environment

Dear Family:

We are about to study the relationships among plants, animals, and the world in which they live. Your entire family might enjoy starting a "Family Ecology Club" and discuss ways in which individuals can help preserve and conserve the planet.

Activities:

- Write a poem about conservation.
- Make up a board game about recycling.
- Make a list of the plants humans need for survival.
- Make a list of the animals humans need for survival.
- Discuss stories in the newspaper about the environment.

Some of the ways I help preserve the earth

List:

1. _____
2. _____
3. _____
4. _____
5. _____
6. _____

Books for Family Reading

The Great Kapok Tree

by Lynn Cherry. A fascinating tale of life and survival in the Amazon Rain Forest.

Global Warming

by Lawrence Pringle. A discussion of the causes and effects of global warming. Presents data on how greenhouse gases can be reduced.

Saving Our Wildlife

by Lawrence Pringle. Points out the relationships necessary between animals and their environment.

Living Things Need Each Other: Ecology & Environment

The biggest challenge for the earth

Three things my family and I do to recycle

What we think should be done about the rain forests of the world

As a family, we believe conservation is . . .

Ways in which our neighborhood works together

Preserving the plants and animals of the world is important because . . .

Things That Matter

Dear Family:

Your child will be reading and studying about matter—what it is made of and how it can be measured. You and your child might want to make a list of the three states of matter (solid, liquid, gas) and how many examples of each can be found in different rooms of your home.

Activities:

- List different ways in which water can be measured.

- Take some photographs of items in your home that represent each of the three states of matter.

- Write a story about "My life as an atom."

- Look for advertisements in the newspaper that use words such as "liquid," "gas," and "solid."

- Talk about all the different ways in which air can be described.

Some of the chemical and physical changes that take place in the kitchen

Books for Family Reading

Atoms, Molecules, and Quarks
by Melvin Berger. A thorough and complete introduction to the tiny particles that make up the basic units of matter.

Gas
by Brian Cook. Gas—how it is discovered, brought to the surface, stored, and used—is presented in this colorful and useful book.

Air and Flight
by Neil Ardley. This book has a number of exciting and fascinating experiments about air that the whole family will enjoy.

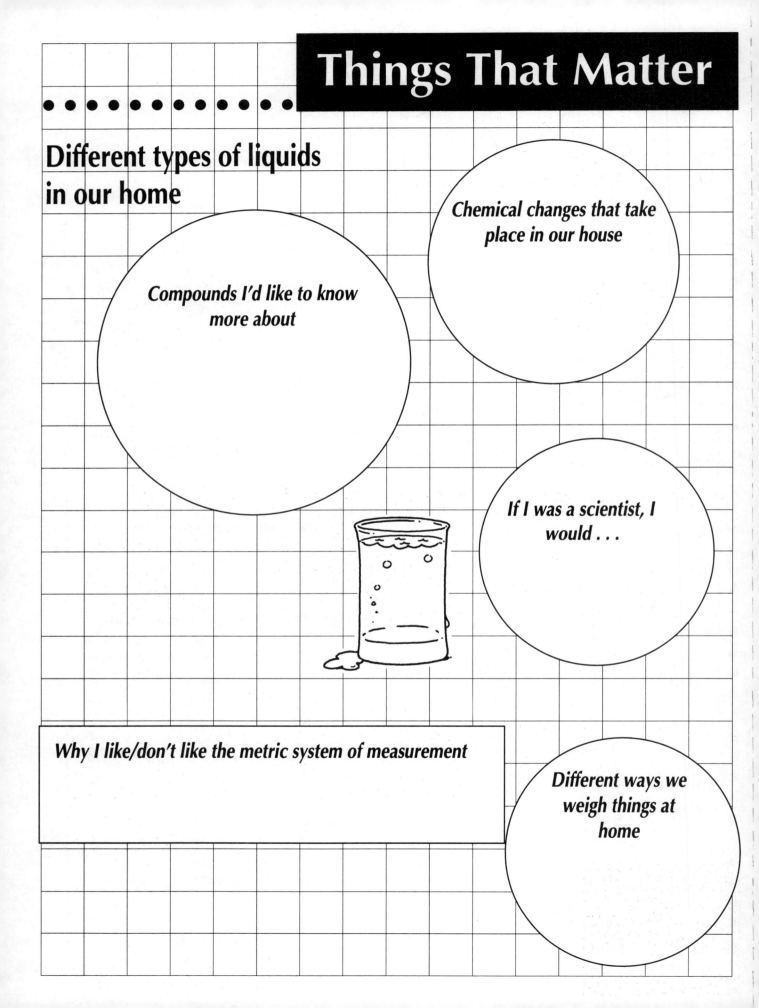

Charge!: Learning About Energy

Dear Family:
Your child will be learning about energy—how it is produced, its different forms, and how it is used. Your entire family might want to put together a list of the different types of energy used in your home.

Activities:

- Make a list of some of the simple machines used in your home.

- List different ways electricity is used in your neighborhood.

- Put together a "Family Safety Guide" for using electrical appliances in the home.

- List different ways light is produced in your home.

- Take photographs of things in your home that produce loud sounds and soft sounds.

Electrical appliances we couldn't live without

Books for Family Reading

Electricity: From Faraday to Solar Generators
by Martin Gutnik. Describes some of the most important inventions and discoveries made throughout history.

The Laser Book
by Clifford Laurence. This is a complete and thorough introduction to lasers and what they are used for.

The Hidden World of Forces
by Jack R. White. In clear language, this book describes force and energy and how they are used.

Charge!: Learning About Energy

- *Ways we can conserve energy in our home*
- *Things that produce sound in my room*
- *Why solar energy is important*
- *Places I have seen rainbows*
- *Different ways in which magnets are used*
- *Kinds of work I like to do*

The Real Dirt: Learning About the Earth

Dear Family:
Your child will be reading and learning about the planet earth—rocks and soil, how the earth moves and changes, and the oceans that cover our planet. Your entire family might enjoy sharing articles from the local newspaper that describe some of the changes that take place on the earth's surface.

Books for Family Reading

Exploring the Sea: Oceanography Today
by Carvel H. Blair. Explores some of the most fascinating aspects of oceans and their inhabitants.

Understanding and Collecting Rocks and Fossils
by Martin Bramwell. A thorough guide to the identification and collection of all kinds of rocks.

Volcanoes: The Fiery Mountains
by Margaret Poynter. Describes in stunning detail the forces that create volcanoes and the damage they do.

Activities:

- Collect different rock samples from around your neighborhood.

- Discuss the dangers of living in areas where earthquakes occur.

- Make a list of some of the most famous volcanoes in the world.

- Plan a voyage across an ocean.

- Take photographs of different land forms in your area.

Places on the earth I would like to visit . . .

The Real Dirt: Learning About the Earth

Ways in which water shapes the land

Ocean creatures I would rather not meet

Ways in which volcanoes are helpful

How people can survive earthquakes

Different ways in which rocks are used in our community

If we could live anywhere, it would be . . .

Weather or Not:
The Forces of Weather and Climate

Dear Family:
Your child will be learning about weather—its different forms, its effects, and how it can be measured. Your family can set up a small weather station outside your house. Weather kits are available in most toy stores.

Activities:

* Share and compare weather predictions on a local TV news program and in your daily newspaper.

* Take photographs of different types of weather and assemble them into a photo album.

* Make a list of your favorite kinds of weather.

* Compare the weather in your area with that of a major city 1000 miles away.

* Write a story about your life as a meteorologist (weather person).

My favorite kind of weather

Books for Family Reading

The Weather Factor
by David Ludlum. A thorough introduction to weather and its effects on humans.

How Did We Find Out About Sunshine?
by Isaac Asimov. A historical overview of discoveries about the sun and its effect on life on earth.

Hurricane
by Faith McNulty. An insightful and fascinating look into hurricanes—how they're formed and the damage they create.

Weather or Not: The Forces of Weather and Climate

What I like to do on rainy days

The kind of weather that scares me the most

Different shapes of clouds I have seen

Trying to predict the weather is like . . .

Where we like to go on sunny days

If I could have one type of weather for a month it would be . . .

Beyond Earth: Exploring the Planets and Space

Dear Family:
Your child will be reading about the solar system, the universe, and the exploration of space. Your family might want to make a list of some of the planets you would like to visit if you could.

Activities:

- Observe the different phases of the moon during one month.

- Write a story about living on a different planet.

- List some of the most important space discoveries.

- Write a letter applying for a job as an astronaut.

- Make a three-dimensional model of the planets of the solar system.

Books for Family Reading

New Worlds: In Search of the Planets
by Heather Cooper and Nigel Henbest. An accurate overview of our current knowledge about each of the nine planets.

Hello? Who's Out There? The Search for Extraterrestrial Life
by Don Dwiggins. A guide to the efforts, real and fanciful, used to locate and substantiate life forms from other worlds.

Galaxies
by Seymour Simon. A wonderful voyage into new worlds and new discoveries far beyond the planet earth.

If I could travel anywhere in the universe, it would be . . .

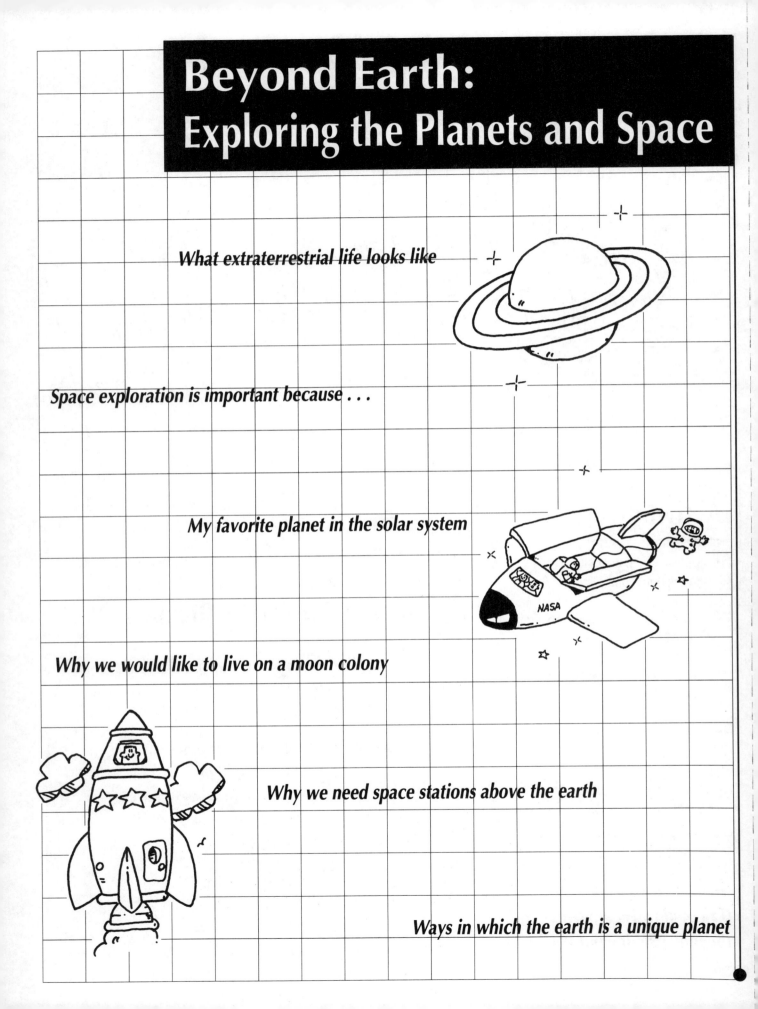

The Body Human: Learning About Ourselves

Dear Family:
Your child will be reading and learning about the human body—how it functions, its various parts, and how we can protect it. Your entire family might want to look through several magazines for various examples of body and health care products advertised.

Activities:

- Make a list of all the diseases family members have had in the past year.

- What is the most important part of your body?

- Write a story from the viewpoint of a body organ, such as the heart, lungs, or stomach.

- Research a deadly disease and how it is being fought.

- Talk to a medical person about proper diets for family members.

How I take care of my body

Books for Family Reading

The Macmillan Book of the Human Body

by Mary Elting. Examines each major part of the human body in exciting and fascinating detail.

Children and the AIDS Virus: A Book for Children, Parents, and Teachers

by Rosmarie Hausherr. A complete and thorough guide to AIDS, presented in a non-threatening and informative format. Perfect for family discussions.

Peak Performance: Sports, Science, and the Body in Action

by Emily Isberg. A revealing introduction to exercise and training and how they influence athletic performance.

The Body Human: Learning About Ourselves

- Ways in which I take care of my eyes

- My favorite body part

- How I help keep all my body systems healthy

- Ways in which the family exercises

- Here's what we think are characteristics of a healthy person:

- If I could look inside my body I would like to see . . .

Special Letters

The projects, activities, and information in this section provide you with the opportunity to share some additional ideas with the parents of your students. Each of the following pages is an extension of one or more of the ideas mentioned in the previous letters. You can use these sheets in several different ways.

1. Duplicate a Special Letter and attach it to one or more Parent Letters during the course of the year. This provides parents with some extended information and activities to share with their children.

2. Send Special Letters home individually with students.

3. Send home a Special Letter instead of one of the Parent Letters with all your students. Do this when you want to emphasize a particular concept or piece of information in the science text.

4. The design and format of the Special Letters allow you to include them as a special page in a school or classroom newsletter regularly sent home to parents.

5. The Special Letters are also appropriate for distribution at Parent-Teacher conferences scheduled throughout the year.

6. Several of the Special Letters can be used repeatedly throughout the year. You can duplicate these Letters and keep them on file to send home periodically.

However you decide to use the Special Letters, they can be an added bonus in helping to establish and maintain positive home-school bonds. Sent home throughout the year, they can help ensure that parents are provided with meaningful and relevant ideas on how to actively participate in their children's science discoveries.

A. Family Science Check-up

The following suggestions give you an opportunity to examine your current family science practices with an eye toward adding more positive science practices to your family's schedule. Begin by placing a check before each true statement. Then place a star by one unchecked item that you and your family would like to try to do. In a month or so, recheck all true statements. Hopefully, the starred item will then become one of the ones that you check.

NOTE: Check only those items that you do on a regular basis.

_____ 1. I share science information from the local newspaper or national magazines with my child regularly.

_____ 2. I buy science-oriented books for birthdays, holidays, or other occasions.

_____ 3. The family goes on "field trips" to various locations in and around the local community and discusses some of the scientific principles we observe.

_____ 4. Family members make regular use of the library to read or consult science-related books.

_____ 5. I encourage my child to talk about some of the new things he or she is learning about the natural world.

_____ 6. My child knows that I have a positive attitude toward science.

_____ 7. I encourage my child to view selected TV programs dealing with the world of science.

_____ 8. My child has an encyclopedia, science dictionary, or other science reference material at home.

_____ 9. I encourage my child to write for science information, catalogs, or materials to use at home.

_____ 10. I talk with my child about some of the new things I am learning about the world.

B. Can You Help Us?

Parents: Please complete the following survey and return it to school at your earliest convenience. Thank you.

Name: _____

Address: _____ Phone number _____

Do you have any special skills that you can share as part of our science program? These may include model building, electrical knowledge, fishing or hunting, cooking, and the like? I'd be willing to share the following _____

Do you collect any special objects that you can share with students? These may include shells, bottles, old photographs, or scientific apparatus. I'd be happy to demonstrate my collection of

_____

Does any part of your work or job involve scientific principles? This may include electricians, mechanics, plumbers, carpenters, machinists, architects, etc. I can share the following parts about my job _____

Do you have an association with any special places we may visit as part of our science program? These may include construction sites, hospitals, television studios, industrial sites, and so on. Yes, I can arrange a visit to _____

Do you know of other individuals in the community (friends, relatives, people at work) who have a special hobby or talent they could share with our class?
Their name _____ Phone number _____

Do you have any special materials at home that we could borrow as part of our science program? These may include antiques, memorabilia, special tools, gadgets, etc.
Yes, I could loan you _____

Is there any other information, materials, places, or data you can share with us that will help our science program? _____

_____

C. Science Magazines for Children

The following list contains the names of some of the more popular magazines for children. You should be able to locate these in your child's school library or your local public library. Consider subscribing to one or more of these so that your child may be able to receive his or her own magazine(s).

Audubon Adventure
National Audubon Society
613 Riversville Rd.
Greenwich, CT 06830
(6 per year)

Chickadee
Young Naturalist Foundation
P.O. Box 11314
Des Moines, IA 50340
(10 per year)

The Curious Naturalist
Massachusetts Audubon Society
Lincoln, MA 01773
(4 per year)

Dolphin Log
Cousteau Society
8430 Santa Monica Blvd.
Los Angeles, CA 90069
(4 per year)

Electric Company
Children's Television Workshop
One Lincoln Plaza
New York, NY 10023
(10 per year)

Exploratorium Magazine
3601 Lyon St.
San Francisco, CA 94123
(4 per year)

Faces
Cobblestone Publishing, Inc.
20 Grove St.
Peterborough, NH 03458
(10 per year)

Junior Astronomer
Benjamin Adelman
4211 Colie Dr.
Silver Springs, MD 20906

Junior Natural History
American Museum of Natural History
New York, NY 10024
(Monthly)

Kind News
The Humane Society of the U.S.
2100 L St., NW
Washington, DC 20037
(5 per year)

My Weekly Reader
American Education Publications
Education Center
Columbus, OH 43216
(Weekly during the school year)

National Geographic World
National Geographic Society
17th and M St., NW
Washington, DC 20036
(Monthly)

Naturescope
National Wildlife Federation
1912 16th St., NW
Washington, DC 20036
(5 per year)

Odyssey
Kalmbach Publishing Co.
1027 North Seventh St.
Milwaukee, WI 53233
(Monthly)

Owl
Young Naturalist Foundation
P.O. Box 11314
Des Moines, IA 50304
(10 per year)

Ranger Rick
National Wildlife Federation
1412 16th St., NW
Washington, DC 20036
(Monthly)

Science World
Scholastic Magazines, Inc.
50 W. 44th St.
New York, NY 10036

Science Weekly
P.O. Box 70154
Washington, DC 20088
(18 per year)

Science News
Science Service, Inc.
1719 N Street, NW
Washington, DC 20036
(Weekly)

Science Activities
4000 Albemarle St., NW
Washington, DC 20016
(4 per year)

Scienceland
Scienceland, Inc.
501 Fifth Ave.
New York, NY 10017
(8 per year)

Space Science
Benjamin Adelman
4211 Colie Dr.
Silver Springs, MD 20906
(Monthly during the school year)

3-2-1 Contact
Children's Television Workshop
P.O. Box 2933
Boulder, CO 80322
(10 issues per year)

Wonderscience
American Chemical Society
P.O. Box 57136 - West End Station
Washington, DC 20037
(4 per year)

Your Big Backyard
National Wildlife Federation
1412 16th St., NW
Washington, DC 20036
(Monthly)

Zoobooks
Wildlife Education, Ltd.
930 West Washington St.
San Diego, CA 92103

D. Science Activity Books

The following books provide you and your child with a wonderful collection of experiments, discoveries, and explorations into all dimensions of science. Available in libraries, book stores, and teacher supply stores, they can offer you a host of exciting ways to learn about science with your child. Check with your school's librarian or local public library for additional resources.

Beastly Neighbors: All About Wild Things in the City, or Why Earwigs make Good Mothers by Mollie Rights. Boston, MA: Little, Brown, 1981.

Children are given opportunities to examine the plant and animal life in their immediate environments. For children who don't believe there is nature in the city, this book is a must.

Blood and Guts: A Working Guide to Your Own Insides by Linda Alison. Boston, MA: Little, Brown, 1976.

Seventy experiments allow children to examine, poke, push, and prod their own anatomies to discover the wonderful "laboratory" they carry with them every day.

The Complete Science Fair Handbook by Anthony D. Fredericks and Isaac Asimov. Glenview, IL: ScottForesman, 1990.

Provides teachers and parents with timetables, project ideas, research sources, and information emphasizing a process approach to the creation of science fair projects. The emphasis is on learning and discovery for all students, not on winning the Grand Prize or a slew of blue ribbons.

Foodworks from the Ontario Science Centre. Reading, MA: Addison-Wesley, 1987.

Provides over 100 science activities and fascinating facts that explore the magic of food. Kids will love this assortment of hands-on activities about a subject they all too often take for granted.

Gee Wiz! by Linda Alison and David Katz. Boston, MA: Little, Brown, 1983.

This book uses children's imaginations to present science as a thinking process. Children are allowed to explore their own interests and make self-initiated discoveries as they learn basic science concepts.

Invention Book by Steven Caney. New York: Workman, 1985.

A wonderful introduction to the inventive process, which shows how children can use the processes of famous (and not so famous) inventors to discover their own natural creativity.

The Kid's Nature Book by Susan Milford. Charlotte, VT: Williamson Publishing, 1989.

If you're looking for an activity-a-day to help your child discover and enjoy the wonders of nature, this is the book! Filled with fascinating facts and a host of investigative activities, this book is an exciting addition to any home library.

Nature Activities for Early Childhood by Janet Nicklesburg. Menlo Park, CA: Addison-Wesley, 1976.

Offers 44 projects which allow young children to observe nature and investigate the wonders of the natural world around them. This hands-on approach to science discoveries will lead to more activities initiated by children themselves.

Raceways: Having Fun with Balls and Tracks by Bernie Zubrowski. New York: Morrow, 1985.

The principles of momentum, acceleration, energy, and gravity are all contained within this little book. Ideas and demonstrations help children question and pursue answers about natural physical forces.

Safe and Simple Electrical Experiments by Rudolph F. Graf. New York: Dover, 1973.

More than 100 electrical experiments are contained within the pages of this book—experiments that use simple and inexpensive materials and rely on hands-on demonstrations for children of all ages.

Science Experiments You Can Eat by Vicki Cobb. New York: Harper & Row, 1972.

Children's natural fascination with food and cooking are the subject of this book, which wonderfully presents science in easy-to-understand and relevant terms. Be sure to check out Cobb's follow-up book, *More Science Experiments You Can Eat.*

Science Brainstretchers by Anthony D. Fredericks. Glenview, IL: ScottForesman, 1991.

Offers more than 70 critical thinking and problem-solving exercises designed to help children use and extend their knowledge in the life, physical, and earth & space sciences.

Science Fare by Wendy Saul with Alan R. Newman. New York: Harper & Row, 1986.

You can't go wrong with this book! It is filled with sources and resources for looking into and learning about science. It's a book you'll turn to time and time again.

The Science Book by Sara Stein. New York: Workman, 1980.

One of the best books around, this volume includes a variety of science experiments and demonstrations that excite children and stimulate families to examine science in all its dimensions—in school, at home, and in the community.

Scienceworks by Ontario Science Centre. Toronto, Canada: Kids Can Press, 1984.

This book takes an open-ended and relaxed approach to science—one that treats children as serious scientists and allows them to take the initiative in a host of scientific investigations. A wealth of science excitement is in these pages.

The Whole Cosmos Catalog of Science Activities by Joe Abruscato and Jack Hassard. Glenview, IL: ScottForesman, 1991.

Every family should have this book. Not only does it contain a wealth of science investigations for kids of all ages, but it is also a delight to read.

E. Sources for Children's Literature

Finding appropriate reading material in science may seem like an overwhelming task. It need not be. There are many resources at your disposal that will be of enormous assistance in identifying and selecting relevant literature. The sources listed below can be found at many public libraries. They offer a wealth of information, annotated bibliographies, and pertinent data on what books to use and how they can be used—not just in the science curriculum, but in all subject areas. Use these resources as well as the knowledge of your school and public librarians to help you find the best books to share with your child.

Publications

Arbuthnot, May Hill. *Children's Books Too Good to Miss*. 8th ed. Cleveland, OH: Press of Case Western Reserve University, 1989.

Barstow, Barbara. *Beyond Picture Books: A Guide to First Readers*. New York: Bowker, 1989.

Children's Books: Awards and Prizes. New York: Children's Book Council, 1981.

The Children's Catalog. New York: H. W. Wilson Co.

Children's Choices. Newark, DE: International Reading Association (issued each year).

Cranciolo, Patricia. *Picture Books for Children*. Chicago: American Library Association, 1990.

Dreyer, Sharon. *The Bookfinder: When Kids Need Books*. Circle Pines, MN: American Guidance Service, 1985.

Eakin, Mary. *Subject Index to Books for Primary Grades*. 3rd ed. Chicago: American Library Association, 1967.

The Elementary School Library Collection. 15th ed. Williamsport, PA: Brodart, 1986.

Ettlinger, John. *Choosing Books for Young People, Volume 2: A Guide to Criticism and Bibliography, 1976-1984*. Phoenix, AZ: Oryx, 1987.

Gillespie, John. *Elementary School Paperback Collection*. Chicago: American Library Association, 1985.

Gillespie, John. *Best Books for Children: Preschool Through Grade Six*. 4th ed. New York: Bowker, 1990.

Hearne, Betsy. *Choosing Books for Children*. New York: Delacorte Press, 1990.

Jett-Simpson, Mary. *Adventuring with Books: A Booklist for Pre-K – Grade 6*. National Council of Teachers of English, 1989.

Kimmel, Margaret M. and Elizabeth Segel. *For Reading Out Loud!* New York: Delacorte, 1988.

Kobrin, Beverly. *Eyeopeners! How to Choose and Use Children's Books About Real People, Places, and Things.* New York: Viking, 1988.

Lima, Carol, and John A. Lima. *A to Zoo: Subject Access to Children's Picture Books.* 3rd ed. New York: R.R. Bowker, 1989.

Lukens, Rebecca. *A Critical Handbook of Children's Literature.* Glenview, IL: ScottForesman, 1986.

Monson, Dianne. *Adventuring with Books: A Booklist for Pre-K – Grade 6.* Urbana, IL: National Council of Teachers of English, 1985.

Norton, Donna E. *Through the Eyes of a Child: An Introduction to Children's Literature.* New York: Merrill Publishing Co., 1991.

Pilla, Marianne L. *The Best: High/Low Books for Reluctant Readers.* Englewood, CO: Libraries Unlimited, 1990.

The New York Times Parent's Guide to the Best Books for Children. New York: Times Books, 1988.

Rollock, Barbara. *The Black Experience in Children's Books.* New York: The New York Public Library, 1984.

Taylor, Barbara M. and Dianne L. Monson. *Reading Together: Helping Children Get a Good Start in Reading.* Glenview, IL: ScottForesman, 1991.

Trelease, Jim. *The New Read Aloud Handbook.* New York: Penguin Books, 1992.

Vandergrift, Kay. *Child and Story: The Literary Connection.* Neal-Schuman, 1980.

F. Activities for Use with Science Books

Library books can provide your child with amazing and exciting journeys throughout the world of science. Making science books a regular part of your reading time together can open up some new worlds of discovery and imagination for your child. Thus I encourage you to ask your child's school librarian or the children's librarian at your local public library for selections on good science books appropriate for your child's age and interests.

Listed below are several extending activities that you and your child can share during the reading of a book or upon completion of a book. Of course, you will not want to attempt all of these for any single book. Discuss two or three of them with your child and select those activities most appropriate to the topic of a particular book and the interest of your child. Please feel free to modify and adapt these activities as you see fit.

1. Have your child write a letter to a friend about what's being learned in a science book.

2. Have your child read and compare several books by the same author.

3. Invite your child to keep a journal or diary about what he or she is learning.

4. Work with your child to make up a mock newspaper about a selected book topic.

5. Set up a "Reading Corner" in your child's room filled with periodicals, books, and other printed materials on a particular topic.

6. You and your child might want to record part of a book on cassette tape.

7. Have your child design a wordless picture book on a book topic.

8. Your child might enjoy creating and producing an original book on a particular topic.

9. Help your child design and write a newspaper article about a particular book.

10. Have your child locate and read a relevant magazine article on a specific topic.

11. Ask your child to write a sequel or prequel to a selected book.

12. Have your child adapt a book into a videotaped news report or TV program.

13. Challenge your child to write a description of a book in 25 words or less. In 50 words or less. In 75 words or less.

14. Ask your child to create interview questions for the author of a book.

15. Work with your child to create a glossary or dictionary of important words in the book.

16. You and your child may want to create word puzzles or crossword puzzles on book information.

17. Direct your child to keep a card file of all the books he or she reads.

18. Invite your child to set up a message center to send messages to family members about information learned.

19. Your child might enjoy creating a calendar of important facts.

20. Have your child put together a scrapbook about important data, information, or facts.

21. Ask your child to write a ten-question quiz for a book read.

22. Encourage your child to interview outside "experts" in the local community.

23. Ask your child to create flash cards using facts from a book.

24. You and your child might want to work together and adapt part of a book into a series of cartoons.

25. Invite your child to illustrate portions of a book.

26. Have your child make an advertisement about a book.

27. Encourage your child to establish a science "museum" in one corner of the living room, applying data in a book or series of books.

28. Ask your child to create a collage on a particular topic from old magazines.

29. Have your child paint a large wall poster about the topic of a book.

30. You and your child might want to design and create a diorama of a natural scene or process.

31. You and your child can create a three dimensional display of artifacts associated with a book.

32. Your child can design clay models of important facts.

33. The family can work together to create a radio show about a book.

34. Challenge your child to create a commercial for a science discovery.

35. Have your child produce a puppet show about part of a book.

G. Book Sharing Questions

As you and your child read a book together, or after you've finished reading a book together, it's important to take a few minutes to talk about some of the information in the book. This should be a pleasurable and enjoyable activity—not a time to "test" how much your child remembered about the book. The intent is to provide you and your child with a comfortable opportunity to discuss common feelings and perceptions about a shared book. By using two or three of the questions below for each book you read, you can help your child appreciate the wonder and joy of science.

1. What did you enjoy most about this book? What did you enjoy least? Can you tell me why?

2. If you were asked to write a new ending for the book, what would you change? Why?

3. What other kinds of events or circumstances could have been included in this book?

4. Why do you think your friends would enjoy reading this book? What would they like most?

5. Are there parts of this book that could be taken out? If so, which ones?

6. Would you want to read this book again? Why?

7. If you could write a letter to the author of this book, what would you want to say?

8. What was something new you learned as a result of reading this book? Did this book have any information that you knew already? If so, what?

9. Would you like to read other books by this author?

10. How is this book similar to or different from other books on the same topic?

H. Science Supply Houses

The following organizations and businesses are good resources for science equipment and supplies. You are encouraged to write to these firms and request copies of their latest catalogs. Besides being good reading, these catalogs offer a wealth of inexpensive and educationally sound materials and supplies for you and your child to enjoy together.

Accent! Science
301 Cass St.
Saginaw, MI 48602
(517) 799-8103

Activity Resources Company, Inc.
P.O. Box 4875
Hayward, CA 94540
(415) 782-1300

Albion Import Export, Inc.
Coolidge Bank Bldg.
65 Main St.
Watertown, MA 02172
(617) 926-7222

American Science Center/Jerryco
601 Linden Pl.
Evanston, IL 60202
(312) 475-8440

Carolina Biological Supply Company
2700 York Road
Burlington, NC 27216
(800) 334-5551

Central Scientific Co.
11222 Melrose Ave.
Franklin Park, IL 60131
(312) 451-0150

Connecticut Valley Biological Supply Co.
Valley Road
P.O. Box 326
Southhampton, MA 01073
(800) 282-7757

Creative Learning Systems, Inc.
9889 Hilbert St., Suite E
San Diego, CA 92131
(619) 566-2880

Creative Learning Press
P.O. Box 320
Mansfield Center, CT 06350
(203) 423-8120

Delta Education, Inc.
P.O. Box 950
Hudson, NH 03051
(800) 442-5444

Denoyer-Geppert Science Company
5711 North Ravenswood Ave.
Chicago, IL 60646
(312) 561-9200

Edmund Scientific
101 E. Gloucester Pike
Barrington, NJ 08007
(800) 222-0224

Educational Activities, Inc.
P.O. Box 392
Freeport, NY 11520
(800) 645-3739

Estes Industries/Hi-Flier
1295 H Street
Penrose, CO 81240
(303) 372-6565

Fisher Scientific Co.
4901 W. Le Moyne St.
Chicago, IL 60651
(312) 378-7770

Frey Scientific
905 Hickory Lane
Mansfield, OH 44905
(419) 589-9905

Hubbard Scientific Co.
P.O. Box 104
Northbrook, IL 60065
(800) 323-8368

Ideal School Supply Co.
11000 S. Lavergne Ave.
Oak Lawn, IL 60453
(312) 425-0800

Learning Things, Inc.
68A Broadway
P.O. Box 436
Arlington, MA 02174
(617) 646-0093

LEGO Systems, Inc.
555 Taylor Road
Enfield, CT 06082
(203) 749-2291

Let's Get Growing
General Seed and Feed Company
1900-B Commercial Way
Santa Cruz, CA 95065
(408) 476-5344

NASCO
901 Janesville Ave.
Fort Atkinson, WI 53538
(414) 563-2446

National Wildlife Federation
8925 Leesburg Pike
Vienna, VA 22180
(703) 790-4000

National Science Teachers Association
1742 Connecticut Avenue, NW
Washington, DC 20009
(202) 328-5800

National Geographic Society
17th and M Streets, NW
Washington, DC 20036
(202) 857-7000

Sargent-Welsh Scientific Co.
7300 N. Linder Ave.
Skokie, IL 60077
(312) 677-0600

Science Kit, Inc.
777 E. Park Drive
Tonowanda, NY 14150
(716) 874-6020

The Science Man
P.O. Box 56036
Harwood Heights, IL 60656
(312) 867-4441

Ward's Natural Science Establishment, Inc.
5100 West Henrietta Road
P.O. Box 92912
Rochester, NY 14692
(716) 359-2502

I. Family Field Trip Sites

Science is all around! You can help your child grow and learn in science when you expose him or her to the wonders of science in everyday life. The list below contains places that you can probably find in and around your local community. (Check the Yellow Pages of your local phone directory under the topics below.) They would be ideal places for the family to visit and examine some of the mysteries and principles which make up the world of science. Take advantage of these inexpensive and educationally valuable sites. Visit them regularly and talk about them with your child.

Factories	Nature Trails
Colleges	Backyard of the school
Recreational places	Sporting events
Airport	Beach
Museums	Farms
Live plays/theater	Aquarium
Zoo	Post Office
Planetariums	Nursing home
Graveyard	Radio/TV stations
Library	Fast food restaurants
Amusement park	Supermarkets
Flea market	Hospital
Fire/police station	Power plants
Shipyard	Ethnic restaurants
Wildlife sanctuaries	Greenhouses
Parks	Hardware stores
Drug stores	Department stores
Festivals	Banks
Printers	Sanitation areas
Recycling center	Cave
Lakes	Symphony/Orchestra
Churches/Synagogues	Fish Hatchery
Mortuary	SPCA
Historical sites	Train station
Airport	Newspaper office

J. Things to Write For

The list below has an assortment of agencies, organizations, and bureaus that offer science-related material for you and your family. I encourage you to write to one or more of these places to obtain interesting and useful ideas for use in your home. Above all, you will be helping your child understand that the world of science is broad and far-reaching—a world that knows no limits.

• Write to The Weather School (5075 Lake Road, Brockport, NY 14420) for their free brochure describing The WeatherCycler Study Kit.

• The National Wildlife Federation has a complete catalog of nature education materials. Write to them at 1400 16th St., NW, Washington, DC 20036-2266 [800-225-5333].

• If you're looking for some software programs on earth science you can get a catalog of more than 70 offerings from RockWare, Inc. (4251 Kipling St., Suite 595, Wheat Ridge, CO 80033 [303-423-5645]).

• You can find a wide assortment of health education materials and publications in a catalog put out by Health Edco (P.O. Box 21207, Waco, TX 76702-9964 [800-299-3366]).

• One of the largest and most complete catalogs of science equipment comes from Edmund Scientific Co. (E900 Edscorp Bldg., Barrington, NJ 08007). If what you're looking for isn't in this catalog, then it probably hasn't been invented yet!

• If you're looking for a complete catalog of model rocketry kits and supplies you can't go wrong with the one put out by Uptown Sales Inc. (33 N. Main St., Chambersburg, PA 17201 [800-548-9941]).

• Kids for Saving Earth (P.O. Box 47247, Plymouth, MN 55447) has a 48-page guidebook entitled "Kids for Saving Earth." Write for information.

• The Environmental Protection Agency (A108 EA, 401 M St. SW, Washington, DC 20460) has two publications you may be interested in—"Books for Young People on Environmental Issues (K-12)" and "Environmental Education Materials for Teachers and Young People (K-12)."

• You can get up-to-date environmental education information through the quarterly publication, *Sierraecology*. Contact The Sierra Club (730 Polk St., San Francisco, CA 94109 [415-776-2211].

• Still looking for environmental materials? Write or call Nature Watch (P.O. Box 1668, Reseda, CA 91337 [818-882-5816]) and ask for their catalog of hands-on games and projects.

• The New York Sea Grant Extension Program (125 Nassau Hall, SUNY, Stony Brook, NY 11794-5002 [516-632-8730]) has a 24-page booklet entitled "Earth Guide: 88 Action Tips for Cleaner Water." Write for a free copy.

- A very interesting catalog of materials and supplies related to astronomy is available from Catalog Request Desk, Astronomical Society of the Pacific (390 Ashton Ave., San Francisco, CA 94112).

- Materials related to earthquakes and earthquake safety can be obtained from National Center for Earthquake Engineering Research (State University of New York at Buffalo, 104 Red Jacket Quadrangle, Buffalo, NY 14261).

- A catalog of environmental education resources can be obtained from the Audubon Naturalist Society (8940 Jones Mill Rd., Chevy Chase, MD 20815).

- Write to the National Gardening Association (180 Flynn Ave., Burlington, VT 05401) for a sample copy of "Growing Ideas, A Journal of Garden-based Learning."

- The Map Distribution Center (U.S. Geological Survey, Federal Center, Box 25286, Denver, CO 80225) has a variety of fascinating and interesting maps available including "Dynamic Planet"—a color map that pinpoints the locations of thousands of volcanoes and earthquakes around the world.

- "Recycling Study Guide" has a host of activities and projects for intermediate students. Contact the Recycling Coordinator, Bureau of Solid Waste Management (Wisconsin Department of Natural Resources, P.O. Box 7921, Madison, WI 53707).

- If you're interested in the greenhouse problem and ways your family can help alleviate it, you'll want to obtain a copy of the booklet "The Greenhouse Crisis: A Citizen's Guide." It's available from the Greenhouse Crisis Foundation (1130 17th St. NW, Suite 630, Washington, DC 20036).

- Write to the Children's Book Council (P.O. Box 706, New York, NY 10276-0706) and ask for a copy of "Choosing a Child's Book." Be sure to include a self-addressed, stamped envelope.

- Write for the latest "Government Books for You" catalog from the U.S. Government Printing Office (Superintendent of Documents, P.O. Box 37000, Washington, DC 20013-7000).

- Write to the Forest Service (U.S. Department of Agriculture, P.O. Box 2417, Washington, DC 20013) and ask for information on obtaining a copy of their poster "How a Tree Grows" (FS-8).

- The American Lung Association (1740 Broadway, New York, NY 10019) has a cartoon story for primary students entitled "Charlie Brown Cleans the Air."

- "A Dictionary of Air Pollution and Hazardous Waste" is a brochure that can be obtained from Air Waste Management Association (P.O. Box 2861, Pittsburgh, PA 15230).

- Contact Keep America Beautiful (99 Park Ave., New York, NY 10016) and ask for "Pollution Pointers for Elementary Students"—a list of environmental improvement activities.

- "About Acid Rain" is a booklet available from Boston Edison (800 Boylston St., Boston, MA 02199).

K. Student Summary Sheet

The Student Summary Sheet is a way for you and your child to record the science books read throughout the year. Your child might want to place this sheet inside a notebook or inexpensive portfolio to track the different types of books he or she obtains from the school library, public library, or local bookstore.

For each book read provide your child with one of the following three options:

READ ALONE: There are many books that your child will want to read on his or her own. For those books read independently, have your child record the title and author and an appropriate check mark in this column.

READ TOGETHER: Occasionally invite your child to share books with other members of the family. This sharing can be as simple as a brief summary given at the dinner table or a conversation between you and your child at some time during the day. Invite your child to create some sort of project (e.g. poster, diorama, mobile, etc.) that displays some of the information in the book. Display the project in your child's room or other prominent location.

WRITE ABOUT: Invite your child to share some of the information in the book with a friend, relative, or neighbor. Your child might write to a selected individual and explain some of the things learned in the book and why that data is important. You may want to have your child dictate a letter to you to be sent to a selected individual.

Student Summary Sheet

Science Books I Have Read, Shared, or Written About

Date	Book Title & Author	Read Alone	Read Together	Write About

L. Science Fair Timetable

Many schools schedule science fairs as part of their overall science programs. These exhibits are valuable opportunities for students to examine and explore an area of science in some detail. However, for many youngsters the project seems overwhelming simply because there seems to be too many things to do at once. The following chart provides you and your child with a workable plan of action that can lead to a successful project. You are encouraged to work with your child in completing this chart.

Date of the science fair   _____
Date to begin working on the project
(count back 12 weeks from the science fair opening date) _____

Scheduled completion date	Actual completion date	
____	____	**Week 1** Choose a topic or problem to investigate. Make a list of resources (school library, community library, places to write, people to interview).
____	____	**Week 2** Select your reading material. Begin preliminary investigations. Write for additional information from business firms, government agencies, and so on. Start a notebook for keeping records. Write down or sketch preliminary designs for your display.
____	____	**Week 3** Complete initial research. Interview experts for more information. Decide how to set up your investigation or experiment. Decide what materials you will use in the display. Create an experimental design.
____	____	**Week 4** Begin organizing and reading the materials sent in response to your letters. Decide whether you need additional material from outside sources. Begin collecting or buying materials for your display. Begin setting up your experiment or demonstration. Add information to project notebook as you get it. Start your collection or experiment.

Scheduled completion date	Actual completion date	
——	——	**Week 5** Learn how to use any apparatus you need. Continue recording notes and observations in your notebook. Set up outline for written report.
——	——	**Week 6** Gather preliminary information in notebook. Work on first draft of written report.
——	——	**Week 7** Start assembling unit display. Continue recording notes. Check books, pamphlets, magazines for additional ideas. Verify information with experts: teachers, professors, scientists, parents.
——	——	**Week 8** Begin designing charts, graphs, or other visual aids for display. Take any photographs you need. Record any observations on experiment. Begin preparing signs, titles, and labels for display unit.
——	——	**Week 9** Have photographs developed and enlarged. Talk with experts again to make sure your work is accurate and on schedule. Begin writing second draft of your report. Continue recording observations in notebook.
——	——	**Week 10** Write text for background of display and plan its layout. Complete graphs, charts, and visual aids. Finish constructing your display. Work on final draft of written report.
——	——	**Week 11** Complete your experiment or collection. Write and type final copy of written report. Letter explanations and mount them on your display. Mount graphs, charts, drawings, photographs. Assemble apparatus or collection items; check against your list.
——	——	**Week 12** Proofread your written report. Set up display at home and check for any flaws (leave standing for two days). Carefully take display apart and transport it to science fair site. Set up display. Check and double-check everything. Congratulate yourself!

M. Safety Rules

The following suggestions are offered as guidelines for parents to follow in using science materials and equipment. You are cautioned that the safety, health, and well-being of your child should be paramount in any scientific investigation or experiment. It is far better to err on the side of caution than it is to place your child in an unsafe situation.

When conducting scientific experiments at home:

1. Do not permit your child to handle science supplies, chemicals, or equipment until he or she has been given specific instruction in their use.

2. Prevent loose clothing and hair from coming into contact with any science supplies, chemicals, equipment, or sources of heat or flame.

3. Instruct your child in the proper use of sharp instruments such as pins, knives, and scissors.

4. Instruct your child never to touch, taste, or inhale unknown substances.

5. Warn your child of the danger in handling hot glassware or other equipment. Be sure insulated mitts and other devices for handling hot objects are available.

6. Check electrical wiring on equipment for frayed insulation, exposed wires, and loose connections.

7. Instruct your child in the proper care and handling of pets, fish, plants, or other live organisms used as part of science activities.

8. Have sufficient lighting to ensure that activities can be conducted safely.

9. Make sure you have a fire extinguisher on hand in any work area.

10. When an activity calls for cutting tough materials, such as bones or heavy plastic, a parent or other adult should do the cutting. Ask children to stand back from the cutting area.

N. Potpourri

The following list contains items, catalogs, information, and assorted resources that you may wish to obtain for use at home. These are presented in no particular order. All the resources help your child understand that science can be found everywhere and that it is a natural and normal part of everyday life.

- If you want to share the "light side" of science with your child, get a copy of *A Treasury of Science Jokes* by Morris Goran (Springfield, IL: Lincoln-Herndon Press, 1986). Filled with hundreds of jokes, puns, and humorous stories (some good, some bad), this collection can be part of your home library.

- The National Wildlife Federation has a complete catalog of nature education materials. Write to them at 1400 16th St., NW, Washington, DC 20036-2266 [800-225-5333].

- If you'd like to find out more about how technology can be used with special education students, contact The Council for Exceptional Children's Center for Special Education Technology. Call them at 800-873-8255 and ask for their "Tech Use Guides."

- One of the most complete catalogs of children's literature is available from Perma-Bound (Vandalia Road, Jacksonville, IL 62650). Be sure to check out their extensive section, "Nonfiction for Children–Science."

- The U.S. Space and Rocket Center in Huntsville, Alabama, offers a number of exciting opportunities for students to study and learn about aerospace technology (you may be familiar with their Space Camp program). Call for information on their "Teaching the Future" program [800-63 SPACE].

- The Environmental Protection Agency (A108 EA, 401 M St. SW, Washington, DC 20460) has two publications you may be interested in—"Books for Young People on Environmental Issues (K-12)" and "Environmental Education Materials for Teachers and Young People (K-12)."

- "To Succeed in Science" is a booklet prepared by the Educational Testing Service (Publication Order Services, P.O. Box 6736, Princeton, NJ 08541-6736). Up to 25 copies can be obtained free of charge.

- Interested in obtaining the Resource Catalog from PBS Video? Call them at 800-424-7963.

- If you're going to teach your child about the environment, you should get a copy of *50 Simple Things Kids Can Do To Save The Earth* by The Earthworks Group (Kansas City, MO: Andrews and McMeel, 1990). Chock full of plans, ideas, and strategies, this is a book you and your child will turn to again and again.

- Modern Talking Picture Service (5000 Park St. N, St. Petersburg, FL 33709) has a video available for loan entitled "Just How Do We Make Electricity?"

- Write or call the Educator's Progress Service, Inc. (214 Center St., Randolph, WI 53956 [414-326-3126]) and ask for current information and prices on the following publications: "The Educator's Guide to Free Science Materials," "Guide to Free Teaching Aids," "Elementary Teachers' Guide to Free Materials," and "Guide to Free Computer Materials."

- If you're looking for magazines for your child, you'll want to get a copy of "Magazines for Children"—a complete directory of children's magazines. Contact the International Reading Association (Order Department, 800 Barksdale Road, Newark, DE 19714).

- A TV series well worth watching is *Discover: The World of Science.* Teaching materials from that PBS series are available from GTE School Program (10 N. Main St., Yardley, PA 19067).

- If you and your family would like to start a wildlife club, you can obtain necessary information from Friends of Wildlife (Box 477, Petaluma, CA 94953). Be sure to include a self-addressed, stamped envelope.

- Contact the American Paper Institute (260 Madison Ave., New York, NY 10016) and request a copy of "How You Can Make Paper."

O. Ten Commitments for Parents

1. I will read various science books and materials with my child(ren) on a regular basis.

2. I will provide my child(ren) with a quiet, comfortable place to read and study.

3. I will encourage my child(ren) to develop a personal library—including science-related books—and will contribute to it regularly.

4. I will provide my child(ren) with a wide range of learning experiences both in and outside of the home.

5. I will talk *with* (not to) my child(ren) on a daily basis.

6. I will praise my child(ren) for at least one success or improvement each day.

7. I will hug my child(ren) at least once a day.

8. I will respect each child as an individual—each with his or her unique talents and abilities.

9. I will provide family activities that encourage my child(ren) to grow in mind, soul, and body.

10. I will encourage my child(ren) to view science as an enjoyable and fulfilling lifetime experience.

_____

Parent Signature

_____

Date

CERTIFICATE OF MERIT

Presented to: _____

For: _____

On this Date: _____

_____
Teacher

_____
Parent